Systems Biology

Topical Conference at the 2011 AIChE Annual Meeting

Minneapolis, Minnesota, USA
16-21 October 2011

ISBN: 978-1-61839-580-1

Printed from e-media with permission by:

Curran Associates, Inc.
57 Morehouse Lane
Red Hook, NY 12571

Some format issues inherent in the e-media version may also appear in this print version.

Printed by Curran Associates, Inc. (2012)

For permission requests, please contact AIChE
at the address below.

AIChE
3 Park Avenue
New York, NY 10016-5991

Phone: (203) 702-7660
Fax: (203) 775-5177

www.aiche.org

Additional copies of this publication are available from:

Curran Associates, Inc.
57 Morehouse Lane
Red Hook, NY 12571 USA
Phone: 845-758-0400
Fax: 845-758-2634
Email: curran@proceedings.com
Web: www.proceedings.com

TABLE OF CONTENTS

Systems Biotechnology: Making Systems Biology Work 1
Sang Yup Lee

Construction of Synthetic Circuits by Harnessing Orthogonal Genetic Parts 2
Tae Seok Moon, Christopher A. Voigt

Construction of a Synthetic Promoter Library Employing Well-Studied Regulatory Components 3
Konstantinos Biliouris, Katherine Volzing, Poonam Srivastava, Yiannis N. Kaznessis

Dynamic Tuning of Glycolytic Flux for Heterologous Production with a 'Glucose Valve' 4
Kevin Solomon, Kristala Prather

Dynamic Metabolic Engineering for Lactate Production 5
Nikolaos Anesiadis, William R. Cluett, Stephen S. Fong, Hideki Kobayashi, Radhakrishnan Mahadevan

Elucidation of Isobutanol Tolerance In Escherichia Coli Through Targeted Genome Engineering 8
Jeremy J. Minty, Jihyang Park, Brian N. Johnson, Ted A. Zaroff, Mark A. Burns, Xiaoxia Nina Lin

Rewiring Intracellular Signaling Pathways to Probe the Robustness of Innate Immune Cell Decision Making 9
Nichole Daringer, Joshua N. Leonard

A Synthetic Ecological System for Pattern Formation 10
Jingjing Sun, Ron Weiss

Lipidomic Profiling and Biomarker Discovery of the Snow Algae Chlamydomonas Nivalis by Salt Stress 11
Na Lv, Dong Wei, Feng Chen

A Systems Approach to Identify Metabolic Pathways Controlling Hepatic Lipotoxicity 13
Alexandra K. Leamy, Jamey Young

Directed Evolution of An E. Coli Methionyl-tRNA Synthetase for Two Strain Cell-Selective Proteomic Labeling 14
Frank Truong, Tae Hyeon Yoo, Thomas J. Lampo, David A. Tirrell

Inference of Predominant Pathways In Cancer: A Proteomic Study Among Three Human Mammary Epithelial Cell Models 15
Yogesh Kulkarni, David Klinke

Metabolomic Analysis of Ovarian Cancer 16
Kathleen Vermeersch, Mark P. Styczynski

Effects of Small-Molecule Inhibition of Central Metabolism in a B-Cell Lymphoma Model of Myc Driven Cancer 17
Taylor A. Murphy, Jamey Young

Mechanochemical Energy Transduction In Muscle Contraction 18
Richard L. Long

Multiscale Variability In Human Endotoxemia: Circadian, Ultradian, and Higher Frequency Rhythms In Heart Rate Variability 19
Jeremy D. Scheff, Steven E. Calvano, Stephen F. Lowry, Ioannis P. Androulakis

Patient-Specific Predictions of Thrombosis 22
Matthew H. Flamm, Tom Colace, Manash Chatterjee, Talid Sinno, Scott L Diamond

A Computational Model of Dynamic Flow In the Respiratory System 24
Aleck Alexopoulos, Costas Kiparissides

Nonlinear Dynamics of Poly(β-hydroxybutyrate) Production In Microorganisms 27
Andre Franz, Hartmut Grammel, Ruxandra Rehner, Achim Kienle

Stochastic Operator Splitting Method for Biological Systems 29
TaiJung Choi, Mano R. Maurya, Daniel M. Tartakovsky, Shankar Subramaniam

How Conjugation Contributes to Antibiotic Resistance of Biofilm - from an Aspect of Population Balance Model with Stochastic Intracellular Gene Regulation 31
Che-Chi Shu, Doraiswami Ramkrishna, Anushree Chatterjee, Wei-Shou Hu

Modeling Circadian Interactions Between Peripheral Clocks and Metabolism 32
Panagiota T. Foteinou, Francis J. Doyle

Metabolite-Centric Approach for Drug Targeting and Discovery Against Microbial Pathogens 35
Hyun Uk Kim, Tae Yong Kim, Sang Yup Lee

Heterogeneous Microbial Populations: Using Flow Cytometric Data for Building Dynamic Distributed Models 36
Rita Lencastre Fernandes, Magnus Carlquist, Luisa Lundin, Anna-Lena Heins, Abhishek Dutta, Ingmar Nopens, Anker D. Jensen, Soren J. Johansen, Anna Eliasson Lantz, Krist V. Gernaey

Efficient Simulation of Dynamic Flux Balance Models ... 38
Stuart Harwood, Kai Hoeffner, Paul Barton
An In-Silico Evolution Algorithm for Obtaining Parameter Values In Stochastic Kinetic Models ... 40
Murali Koneru, Michael Benton, Martin A. Hjortsø
Spatial Mechanisms for the Stepwise Navigation of Neutrophils ... 41
Yuki Kimura, Ashish Kapoor, Paul J. A. Kenis, Luke Olson, Christopher V. Rao
Dynamic Metabolic Flux Analysis (DMFA): A Framework for Determining Fluxes At Metabolic Non-Steady State ... 42
Maciek R. Antoniewicz, Robert Leighty
Genome Scale Metabolic Reconstruction and Hypothesis Testing In the Methanogenic Archaeon Methanosarcina Acetivorans C2A ... 43
Matthew Nicholas Benedict, Matthew C. Gonnerman, William W. Metcalf, Nathan D. Price
Proteomic, Physiological, and In Silico Testing of Economic Tradeoffs In Metabolic Networks ... 44
Ross P. Carlson, Reed Taffs, James P. Folsom
Pattern Formation In the Acto-Myosin Cortex: An Active Reaction-Diffusion-Advection Process ... 45
Justin S. Bois, Frank Julicher, Stephan W. Grill
SHREDing a Biochemical Network Into Hierarchical Modules ... 46
Gautham V. Sridharan, Soha Hassoun, Kyongbum Lee
Shewanella Oneidensis MR-1 Population-Level Chemotaxis In a Diffusion Gradient Chamber ... 48
Rui Li, Jennifer M. Auchtung, James M. Tiedje, R. Mark Worden
Higher Order Gene Deletion Mutants Give Insights On Latent Pathway Activation ... 49
Pratish Gawand, Radhakrishnan Mahadevan
Derivation of a Kinematic Model Coupling Stress Fiber Dynamics with JNK Activation In Response to Matrix Stretching ... 51
Roland Kaunas, Zuyi Huang, Juergen Hahn
A Systems Biology Approach to Identify Immune Targets That Control Tuberculosis Granuloma Function ... 54
Mohammad Fallahi-Sichani, Denise E. Kirschner, Jennifer J. Linderman
Dissecting Gene Regulation Through DNA Loop Formation Using Statistical Mechanical Models and Experiments ... 55
James Boedicker, Hernan Garcia, Rob Phillips
Regulation of Epidermal Growth Factor Receptor Phosphorylation and Endocytosis Dynamics by Protein Tyrosine Phosphatases ... 57
Calixte S. Monast, Matthew J. Lazzara
Identifying Perturbed Pathways In Glioblastoma Through Network Reconstruction and Analysis ... 59
James A. Eddy, Nathan D. Price
Peripheral Clock Gene Entrainment by Cortisol ... 61
Panteleimon D. Mavroudis, Jeremy D. Scheff, Steven E. Calvano, Stephen F. Lowry, Ioannis P. Androulakis
Towards Genome-Scale Dynamic Modeling of Cellular Metabolism. the Cybernetic Approach ... 65
Hyun-Seob Song, Doraiswami Ramkrishna
A General Framework for Modeling Growth and Division of Mammalian Cells ... 67
John H. Gauthier, Phillip I. Pohl
Phenotype Prediction Through Genome-Scale Modeling and Raman Spectroscopy ... 68
Ryan S. Senger
Quantitative Model Analysis with Qualitative Biological Data: Stem Cell Autoregulation Mechanisms In the Drosophila Germarium ... 69
Michael S. Pargett, Robin Harris, Hilary Ashe, David M. Umulis
Dynamic Response of C. Elegans Reproduction to Chronic Heat Stress ... 70
Peter B. Winter, Patrick D. McMullen, Erin Aprison, Luis A. N. Amaral, Richard I. Morimoto, Ilya Ruvinsky
Metabolic Network Reconstruction and Genome-Scale Model of Butanol-Producing Strain Clostridium Beijerinckii NCIMB 8052 ... 72
Caroline B. Milne, Nathan D. Price
A Novel FBA Strategy for Metabolic Modeling of Cyanobacteria ... 74
Tirthankar Sengupta, Mani Bhushan, Pramod P. Wangikar
Constraint-Based Modeling of Cyanobacterial Metabolism ... 76
Trang T. Vu, Sergey M. Stolyar, Grigoriy E. Pinchuk, Allan E. Konopka, Alex S. Beliaev, Jennifer L. Reed
Metagenome Based Metabolic Network Analysis of Human Microbiome ... 78
Yu Chen, Xiaoxia Nina Lin
In Silico Metabolic Model of Rhodococcus Erythropolis to Study and Improve Desulfurization ... 79
Shilpi Aggarwal, Iftekar A. Karimi, Dong-Yup Lee
Contrasting the Metabolic Capabilities of Multiple Cyanobacterial Species ... 82
Rajib Saha, Bertram M. Berla, Thanura Elvitigala, Lawrence E. Page, Himadri B. Pakrasi, Costas D. Maranas

Genome Scale Analysis and Dynamic Study of Methanogen, Methanosarcina Barkeri ... 83
Matthew C. Gonnerman, Nathan D. Price
Predicting Evolution ... 84
Pornkamol Unrean, Pedro V. Pena, Friedrich Srienc
The Antibiotic Resistance Game: The Genetic Regulation of Antibiotic Resistance Transfer In Enterococcus Faecalis ... 85
Anushree Chatterjee, Che-Chi Shu, Yiannis Kaznessis, Doraiswami Ramkrishna, Gary M. Dunny, Wei-Shou Hu
A Novel Multi-Layer Inference Approach to Reconstruct Conditional-Specific Gene Network ... 86
Ming Wu, Christina Chan
Regularization of Ill-Conditioned Inverse Problems Resulting From Determining Transcription Factor Profiles From Fluorescent Reporter Images ... 87
Loveleena Bansal, Yunfei Chu, Carl D. Laird, Juergen Hahn
Systems Modeling of Epidermal Growth Factor and Nerve Growth Factor Signaling: New Insights Into Transient Vs. Sustained ERK Activation Controlling Cell Fate Decisions ... 90
Marc R. Birtwistle, Alexander von Kriegsheim, Walter Kolch
A Novel Mathematical Model of TGF-β-Induced Apoptosis Signaling Pathways ... 91
Seung-Wook Chung, Carlton R. Cooper, Mary C. Farach-Carson, Babatunde A. Ogunnaike
Feedback Effects of Extracellular Matrix Composition On AMPK Signaling Pathway: Pathway to Novel Drug Therapeutics Targeting Atherosclerosis ... 93
Prashanthi Vandrangi, John J. Y. Shyy, Victor G. J. Rodgers
Genome Alignment and Pair-Wise Comparison of Microbial Metabolic Networks Identifies Novel Potential Drug Targets ... 95
Joshua J. Hamilton, Jennifer L. Reed
Insights From High-Throughput Reconstruction and Analysis of 3500 Genome-Scale Metabolic Models ... 97
Christopher Henry, Sam Seaver, Fangfang Xia, Scott Devoid, Matt DeJongh, Aaron Best, Veronika Vonstein, Ross Overbeek, Rick Stevens
Linking Genotype and Phenotype Through a Graded Dominant Mutant Phenotype Approach ... 98
Amanda M. Lanza, John Blazeck, Nathan Crook, Hal Alper
Comparative Transcriptome Analysis In Baby Hamster Kidney Cells Through High Throughput Sequencing ... 99
Nitya M. Jacob, Kathryn C. Johnson, Nandita Vishwanathan, Karthik P. Jayapal, Chetan T. Goudar, Wei-Shou Hu
Transcription-Based Functional Modeling of the Host Response to Influenza Virus Infection: Defining a Healthy Immune Response ... 100
Jason E. Shoemaker, Satoshi Fukuyama, Saori Sakabe, Hiroaki Kitano, Yoshihiro Kawaoka
Genome-Level Quantification of the Population Dynamics of Virus and Virus-Like Defective Interfering Particles ... 103
Collin M. Timm, Fulya Akpinar, John Yin
Identification of a Novel Target for Breast Cancer by Exploring Gene Switches On a Genome Scale ... 104
Ming Wu, Li Liu, Christina Chan
Short-Term Hepatic Gene Expression Profiling Following Thermal Injury ... 105
Qian Yang, Mehmet A. Orman, Francois Berthiaume, Marianthi G. Ierapetritou, Ioannis P. Androulakis
Author Index

SYSTEMS BIOTECHNOLOGY: MAKING SYSTEMS BIOLOGY WORK

Professor Sang Yup Lee, KAIST, Daejeon, Republic of Korea

Abstract

Systems biology has been changing the way biological systems are studied. Its capability to decipher the biological system at a much larger scale is transforming the way biotechnological research is performed as well. We are now observing the paradigm change in biological and biotechnological research on microorganisms as systems biology and synthetic biology are playing increasingly important roles in deciphering and redesigning the cellular networks. In this lecture, I will present the general strategies for systems biotechnology, which will be accompanied by several successful application examples including production of chemicals and materials. Also, a general strategy for systems metabolic engineering will be described as an essential technology for making any bioprocess competitive. [This work was supported by the Korean Systems Biology Research Grant and the Global Frontier Research Program from the Korean Ministry of Education, Science and Technology.]

(126a) Construction of Synthetic Circuits by Harnessing Orthogonal Genetic Parts

Tae Seok Moon and Christopher A. Voigt, Department of Pharmaceutical Chemistry, University of California, San Francisco, San Francisco, CA

Many applications require cells to integrate multiple environmental signals and to implement synthetic control over biological processes. Genetic circuits enable cells to perform computational operations, interfacing biosensor modules and actuator modules. Despite advances in the rational construction of artificial biological systems, synthetic biologists are still limited by the lack of well-characterized orthogonal parts and predictive computational models. Here, we harness a toolbox of genetic parts from bacteria to construct a genetic multiplexer, three 2-input AND gates, a 3-input AND gate, and a 4-input AND gate. To this end, genetic parts have been screened and modified to make them orthogonal, a transfer function model has been developed, and such well-characterized modules have been rationally combined. This work represents a step toward constructing programmed cells that are able to process multiple input signals and to produce desirable outputs for real-world applications.

See more of this Session: Synthetic Systems Biology
See more of this Group/Topical: Topical A: Systems Biology

(126b) Construction of a Synthetic Promoter Library Employing Well-Studied Regulatory Components

Konstantinos Biliouris[1], Katherine Volzing[1], Poonam Srivastava[2] and Yiannis N. Kaznessis[1], (1) Chemical Engineering and Materials Science, University of Minnesota, Minneapolis, MN, (2) Biotechnology Institute, University of Minnesota, St Paul, MN

Even though the evolution of synthetic biology over the last decade has been signicant, the number of molecular parts with well-defined functions still remains small.The need for developing libraries of well-characterized molecular components has been discussed before [1]. Libraries of well-defined molecular elements will allow to program biological function easier and in a more predictable manner. To this end, we develop a library of synthetic inducible promoters employing regulatory elements of the well-studied tetracycline, lactose and lambda-phage operons. We demonstrate that the dierential positioning of operator sites yields a range of expression and repression proles.We tested the promoters strength and leakiness using flow cytometry and monitoring the output signal for a wide range of inputs. We next explored the modularity of the designed promoters. Lastly, we introduced point mutations to two of the synthetic promoters and analyzed the role of individual nucleotides on the overall promoter behavior. In conjunction with the experiments, we developed mathematical models that capture the behavior of the synthetic promoters. This study provides a dozen varying strength synthetic promoters that can be readily utilized as stand-alone molecular systems or integrated into higher level synthetic devices.

References

1) T. Ellis, X. Wang, and J.J. Collins. Diversity-based, model-guided construction of synthetic gene networks with predicted functions. Nature biotechnology, 27(5):465-471, 2009.

See more of this Session: Synthetic Systems Biology
See more of this Group/Topical: Topical A: Systems Biology

(126c) Dynamic Tuning of Glycolytic Flux for Heterologous Production with a 'Glucose Valve'

Kevin Solomon and Kristala Prather, Chemical Engineering, Massachusetts Institute of Technology, Cambridge, MA

With the rising costs of petrochemical feedstocks and growing concern regarding the sustainability of chemical synthesis, microbial production has become an increasingly attractive alternative to the production of commodity and specialty chemicals. Great strides have been made with the development of existing and novel chemistries to generate a wide array of compounds; however, efforts to make these systems economically viable still face a number of challenges. Notable among these is the issue of pathway flux manipulation. In this study, we examine this problem as it pertains to the redirection of glucose from glycolysis.

We have been able to create a new *E. coli* platform that decouples glucose transport and phosphorylation allowing us to independently control the flux of glucose and enable the construction of pathways requiring unphosphorylated glucose directly. Through the use of transcriptional and post transcriptional devices such as antisense RNA, inverting gene circuits and constitutive promoters, we explore a wide range of glycolytic flux and study its impact on both heterologous pathway productivity and cellular processes. In this manner, we are able to increase molar yield in a model glucose consuming pathway by at least 50%. These devices also allow for novel dynamic optimization schemes, pathway development and perhaps improvement of existing processes. In this talk, I will highlight some of the successes and challenges associated with flux manipulation as well as future areas of development.

See more of this Session: Synthetic Systems Biology
See more of this Group/Topical: Topical A: Systems Biology

DYNAMIC METABOLIC ENGINEERING FOR LACTATE PRODUCTION

N. Anesiadis, University of Toronto, Toronto, Ontario, Canada

W. R. Cluett, University of Toronto, Toronto, Ontario, Canada

H. Kobayashi, Japan Agency for Marine-Earth Science and Technology, Japan

S. Fong, Virginia Commonwealth University, Richmond, VA

R. Mahadevan, University of Toronto, Toronto, Ontario, Canada

The production of chemicals in biological hosts has attracted significant academic and commercial interest in the past decades due to fossil fuels depletion and the necessity for a sustainable planet. Research has focused not only on low cost fuels, but also on value-added chemicals. Although there are many microorganisms naturally producing biochemicals, genetic engineering is typically applied to improve their performance.

In metabolic engineering, the target is to eliminate competing by-products in order to increase the yield of the desired product. However, this increase usually comes at the expense of reduced growth rate and in some cases "sick" strains grow very poorly. In addition, expression of non-native pathways in microbial hosts is commonly used in the production of fuels, vitamins, proteins and pharmaceuticals. However, the cost of inducers for non-native pathways in large-scale processes can be significant and elimination of this step could be crucial for economic viability. The work presented here shows some preliminary results towards the optimization of a bioprocess when growth impairment is the bottleneck of the designed strain.

The production of chemicals as the outcome of metabolism is in conflict with the organism's natural tendency for growth. This trade-off between growth and product formation requires further optimization to improve bioprocess economics.

Ideally, it is optimal to maximize the biomass formation in the first phase of a batch to achieve high density of the biocatalyst as fast as possible. Then, in the second phase, most of the carbon can be rerouted towards the desired product. These distinct phases have a significant impact on the productivity and yield of a process. Productivity depends on the amount of catalyst (i.e. biomass) and is optimized in the first phase. High yield is achieved by deleting pathways leading to competing by-products in the second phase.

In order to achieve this scheme, we proposed the dynamic control of gene expression for those pathways contributing to growth at the expense of byproducts formation (Anesiadis *et al.* 2008). Dynamic control was applied by placing these genes under the control of the toggle switch, a bistable genetic controller (Gardner *et al.* 2000 and Kobayashi *et al.* 2004). Genes supporting growth are expressed initially leading to a fast growth phase, followed by repression of these genes leading to a high production phase.

Gene repression is IPTG-inducible, however the ultimate goal is to couple the genetic controller with a quorum sensing system and tune the genetic circuit in order to apply repression of the genes at the optimal time. The programming and tuning of such a process is expected to eliminate the need for inducer, thus decreasing the operational cost.

Here, we evaluate the dynamic method for the production of lactic acid in the MG1655 Δ(*adhE pta*) mutant under anaerobic conditions (Fong *et al.* 2005). The strain was transformed with plasmid pTOG(*pta*) carrying the *pta* gene on the genetic toggle switch background. A mutant carrying plasmid pTOG(*gfp*) served as the control. Expression of *pta* and *gfp* was induced by heat shock at 42°C for 30 min at the beginning of the fermentation. Repression was achieved by the addition of isopropyl-β-D-thiogalactopyranoside (IPTG) to a concentration of 2 mM. The strains were grown in mineral medium supplemented with 2 g/L glucose, 0.5 g/L yeast extract and 100 mg/L ampicillin in all stages of the inoculum preparation and the final characterization. Optical density was measured at 600 nm in a Jenway 6320D spectrometer. Glucose and lactate were measured with a Shodex RI detector using an Aminex HPX-87X ion exchange column at 25°C with 0.5 and 8.5 mM H_2SO_4 as the mobile phase, respectively.

Time profiles of optical density and lactate concentration were obtained for triplicate experiments and one standard deviation was considered (Figure 1). IPTG was added after 5 hr.

The cell density confirms that the cells expressing *pta* grow faster than the control cells carrying the pTOG(gfp) plasmid. As a result of higher growth rate, lactate was produced at a higher titer. The final lactate concentration of the dynamic strategy is approximately 10% higher than the control (14.9 compared to 13.3 mM of lactate at the end of the fermentation).

These results show the proof of concept proposed in our earlier paper (Anesiadis *et al.* 2008). However, optimization of this method is required. The repression time (i.e. the time of IPTG addition) is one parameter that could be optimized. The next challenging step is to couple the pTOG(pta) plasmid with a quorum sensing system in order to repress *pta* expression in an inducer-free process.

The novelty of the method arises from the dynamic expression of otherwise deleted genes that contribute to growth rate and productivity. This method is expected to be significant for improving the applicability of engineered strains that have low growth rate.

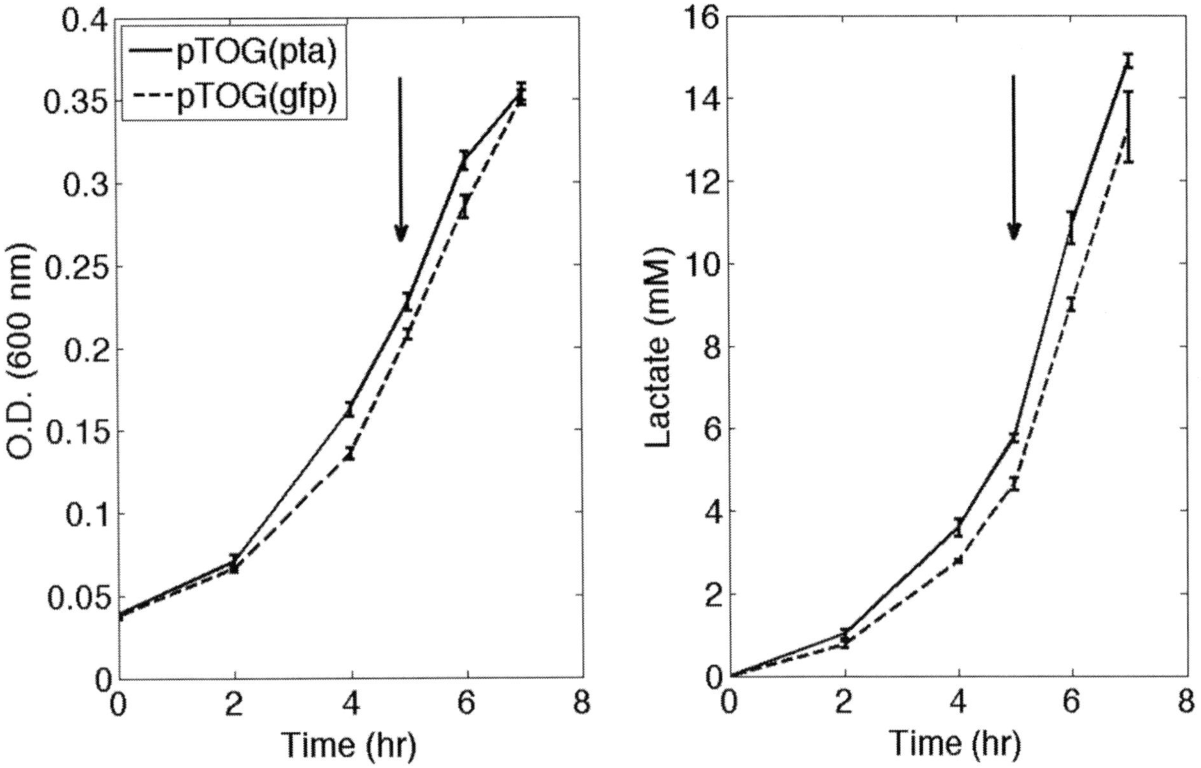

Figure 1: Time profiles of optical density and lactate concentration

References:

1. Anesiadis N., Cluett W.R. & Mahadevan R. (2008) Dynamic metabolic engineering for increasing bioprocess productivity. *Metab. Eng.* 10: 255-266.

2. Fong S., Brugard A.P., Herring C.D., Knight E.M., Blattner F.R., Maranas, C.D. & Palsson B.O. (2005) *In silico* design and adaptive evolution of *Escherichia coli* for production of lactic acid. *Biotechnol. Bioeng.* 91: 643-648.

3. Gardner T.S., Cantor C.R. & Collins J.J. (2000) Construction of a genetic toggle switch in *Escherichia coli. Nature.* 403: 339-342.

4. Kobayashi H., Kærn M., Araki M., Chung K., Gardner T.S., Cantor C.R. & Collins J.J. (2004) *Proc. Natl. Acad. Sci.* 101: 8414-8419.

(126e) Elucidation of Isobutanol Tolerance In Escherichia Coli Through Targeted Genome Engineering

Jeremy J. Minty, Jihyang Park, Brian N. Johnson, Ted A. Zaroff III, Mark A. Burns and Xiaoxia Nina Lin, Department of Chemical Engineering, University of Michigan, Ann Arbor, MI

Advances in microbial engineering have led to the development of metabolic pathways for producing higher molecular weight alcohols as next-generation biofuels. However these molecules are highly toxic to microbes, which may limit product titer and reduce volumetric productivity [1]. Due to broad mechanisms of toxicity, tolerance to alcohols and other solvents is a complex trait that involves a diversity of cellular adaptations and responses that probably contribute synergistically to the overall phenotype [2]. Guided by evolutionary-genomic studies [3], we are using Multiplex Automated Genome Engineering (MAGE) [4] to generate combinatorial libraries of 35 mutations associated with isobutanol tolerance in *E. coli*. Variants with improved isobutanol tolerance are isolated using high-throughput phenotype screening with a microfluidic platform and then further characterized by precisely determining isobutanol tolerance, isobutanol productivity, and genotype. This approach allows systematic correlation of isobutanol tolerance phenotypes and genotypes, yielding insights into mechanisms of tolerance, as well as generating improved strains of *E. coli* that may be immediately useful in the production of isobutanol.

[1] Atsumi, Shota, Taizo Hanai, and James C. Liao. "Non-Fermentative Pathways for Synthesis of Branched-Chain Higher Alcohols as Biofuels." Nature 451 (2008): 86-91.

[2] Nicolaou, Sergios A., Stefan M. Gaida, and Eleftherios T. Papoutsakis. "A Comparative View of Metabolite and Substrate Stress and Tolerance in Microbial Bioprocessing: From Biofuels and Chemicals, to Biocatalysis and Bioremediation." Metabolic Engineering 12 (2010): 307-31.

[3] Minty, Jeremy J., Ann A. Lesnefsky, Fengming Lin, Yu Chen, Ted A. Zaroff, Artur B. Veloso, Bin Xie, Catie A. McConnel, Rebecca J. Ward, Donald R. Schwartz, Jean-Maire, Rouillard, Yuan Gao, Erdogan Gulari, and Xiaoxia Nina Lin. "Evolution combined with genomic study elucidates genetic bases of isobutanol tolerance in Escherichia coli." Microbial Cell Factories (2011) 10:18.

[4] Wang, Harris H., Farren J. Isaacs, Peter A. Carr, Zachary Z. Sun, George Xu, Craig R. Forest, and George M. Church. "Programming Cells by Multiplex Genome Engineering and Accelerated Evolution." Nature 460 (2009): 894-99

See more of this Session: Synthetic Systems Biology
See more of this Group/Topical: Topical A: Systems Biology

(126f) Rewiring Intracellular Signaling Pathways to Probe the Robustness of Innate Immune Cell Decision Making

Nichole Daringer and Joshua N. Leonard, Chemical and Biological Engineering, Northwestern University, Evanston, IL

The goal of this project is to investigate how innate immune cells process disparate extracellular stimuli via intracellular signaling pathways to arrive at a coherent response and determine whether signal rewiring can affect this response. In particular, innate immune cells such as dendritic cells and macrophages may become functionally polarized towards either immunostimulatory or immunosuppressive phenotypes, a process that plays an important role in controlling local immune responses during many chronic diseases. However, the regulation of this decision point is poorly understood. In the case of cancer, signals within the tumor microenvironment induce local immune dysfunction, which causes responding immune cells to promote tumor growth rather than tumor destruction and clearance. It has been reported that the influence of such immunosuppressive stimuli can be overridden by intratumoral injection of immunostimulatory ligands such as Interleukin 12 (IL-12) or agonists for the Toll-like Receptors (TLRs), which induce potent immune activation in response to pathogen associated molecules.

In order to probe the robustness of innate immune cell decision making in response to contradictory intracellular signaling, we constructed a library of receptors that rewire various polarization-related intracellular signaling pathways in macrophages and dendritic cells. We hypothesize that some types of rewiring may be sufficient to fully invert the response of a cell to a given microenvironment. In particular, we sought to identify a rewiring strategy that would invert the response of an innate immune cell to a tumor-associated immune microenvironment, since such an approach may be of therapeutic relevance to the treatment of cancer. Top-down and bottom-up rewiring strategies were included in the construction of this library, meaning that rewiring was accomplished both through the use of chimeric receptors that integrate with native signaling pathways and orthogonal sensors and signaling pathways. We will describe the use of this library to probe the robustness of innate immune cell polarization and re-polarization in tumor-associated immune microenvironments.

See more of this Session: Synthetic Systems Biology
See more of this Group/Topical: Topical A: Systems Biology

(126g) A Synthetic Ecological System for Pattern Formation

Jingjing Sun, Dep. of Biological Engineering, M.I.T., Cambridge, MA and Ron Weiss, Department of Biological Engineering, M.I.T., Cambridge, MA

Synthetic biology is the design and construction of new biological entities such as proteins, genetic circuits, and cells or even the redesign of existing biological systems. In my work, I design and construct a synthetic ecological system to study how simple local rules lead to global pattern formation.
In nature, programmed cell death is often used to remove excess cells, as is seen in embryo, finger, and brain development, sea urchin metamorphosis, and plant immune responses. Within the context of synthetic biology, coordinating cell death and survival artificially will enhance our understanding of natural systems and our ability to engineer complex cellular patterns. In addition, the ability to precisely regulate cell density is essential for numerous medical and industrial applications. To explore coordination of cell death and survival and pattern formation, I implemented a synthetic ecological system with the following game rules: 1) cells die when they are in an overpopulated environment due to overpopulation (too many neighbors); 2) cells die when they are in solitude (too few neighbors); 3) cells survive and proliferate only when there are an intermediate number of neighbors. By changing the underlying local interaction rules, interesting 2D global patterns emerge after repeated rounds of cell death and survival. This work provides an important platform for understanding how local rules dictate complex global behavior.
In order to achieve this goal, I implement a genetic circuit with three modules into yeast. 1) The first is an artificial cell-to-cell communication module that detects neighboring cell density using a small diffusible signal molecule; 2) the second is a low threshold killing and rescuing module that kills cells at low signal levels but prevents cell death at high signal levels; and 3) the third is a high threshold killing module that kills cells at high levels of signal molecules.

See more of this Session: Synthetic Systems Biology
See more of this Group/Topical: Topical A: Systems Biology

(182a) Lipidomic Profiling and Biomarker Discovery of the Snow Algae Chlamydomonas Nivalis by Salt Stress

Na Lv[1], Dong Wei[1] and Feng Chen[2], (1)School of Light Industry and Food Sciences, South China University of Technology, Guangzhou, China, (2)Institute for Food & Bioresource Engineering, College of Engineering, Peking University, Beijing, China

Microalgae can respond quickly to the stress environments through a variety of molecular mechanisms for survival. Salinity is one of the most important factors in these harsh environments. Under extremely high salinity conditions, microalgae show various mechanisms, such as reducing osmotic stresses, protecting chloroplast functions and/or maintaining ion homeostasis, etc. The regulation of lipid biosynthesis in microalgae is considered a crucial pathway in resistant physiology to stress stimulus which, however, is still not fully understood.

In the present work, *Chlamydomonas nivalis*, the snow alga which is found in polar region and similar extreme environments, was used to as a model organism to investigate the mechanism of cell response and adaptation to salt stress on the lipidomic level. The 6-day culture of *C. nivalis* UTEX LB 2824 in Bold 1NV medium under autotrophic growth was stressed by NaCl with final concentrations of 0.25%, 0.50%, 1.00% for 1, 7, 15 h, respectively. Ten mg of dried biomass from each stressed culture was extracted with $CHCl_3$:CH_3OH:H_2O containing BHT to obtain the total lipids. All lipid samples were analyzed by a novel approach of UPLC/Q-TOF-MS-based lipidomic profiling coupled with multivariate statistical analysis for lipid biomarker discovery. 4,975 peaks of ions in ESI positive mode and 1,016 peaks of ions in ESI negative mode were presented in Based Peak Intensity (BPI) chromatograms. The data of both ESI modes were normalized and pareto-scaled followed by feeding to SIMCA-P$^+$ V12.0 software to perform PLS-DA models, respectively. The samples were basically classified into three catalogues in PLS-DA score plots: 0.25%-NaCl/control, 0.5%-NaCl and 1%-NaCl group. 0.25%-NaCl group has insignificant difference with the control on lipid profile, but significantly different with 0.5%-NaCl and 1%-NaCl group. The time of NaCl stress had little effect on lipid profile. Three pairs of groups, control/0.5%-NaCl, control/1.0%-NaCl and 0.5%-NaCl/1.0%-NaCl, were analyzed separately by OPLS-DA models to select potential biomarkers. Twenty-six ions in ESI positive mode and nine ions in ESI negative mode were finally selected as lipid biomarkers and then identified by resolving the structure. All biomarkers are polar lipids and belong to seven types and thirty-five kinds, including three of MGDG, five of DGDG, sixteen of DGTS, two of PG, four of SQDG, four of PE and one of PI, which are in relation to the regulation of cell membrane stability, signal transduction and photosynthesis efficiency. These results demonstrated *C. nivalis* had developed a plethora of biochemical mechanisms by evolution to cope with salt stress: to synthesis new lipids from starting, digest the existing lipids completely or adjust their quantities and composition. This was the first report to demonstrate the lipid profile in microalgae corresponding to stress conditions by "lipidomics" technology, and the discovered lipid biomarkers with up- or down-regulation provided new insights into the explanation of the complex response pathways of microalgae to stress conditions on the lipidomic level.

See more of this Session: Proteomics & Metabolomic Approaches to Systems Biology
See more of this Group/Topical: Topical A: Systems Biology

(182b) A Systems Approach to Identify Metabolic Pathways Controlling Hepatic Lipotoxicity

Alexandra K. Leamy, Chemical & Biomolecular Engineering, Vanderbilt University, Nashville, TN and Jamey Young, Chemical and Biomolecular Engineering, Vanderbilt University, Nashville, TN

In the past half century, the prevalence of diabetes and obesity has grown significantly, a trend which has been described as a health epidemic in the US. The similar pathogenesis of these two diseases involves abnormal accumulation of lipids (steatosis) in non-adipose tissues due to elevated free fatty acids. In the liver, saturated fatty acids (SFA) overexposure has been shown to induce ER stress and oxidative stress in hepatocytes, leading to increased apoptosis. However, the exact identity of the lipid species responsible for these effects is still largely uncertain, and the mechanism by which ER and redox homeostasis becomes dysregulated remains unknown.

In order to better characterize the phenotypic response of hepatic cells to SFA overexposure, we have sought to systematically perturb the main routes of fatty acid utilization and disposal, including β-oxidation, triglyceride (TG) synthesis and phosphotidylcholine (PC) synthesis. Experimentally, H4IIEC3 hepatic cells are incubated with varying concentrations of SFA (palmitate, 200-400 μM) in combination with treatments designed to prevent reactive oxygen species (ROS) accumulation and lipoapoptosis by modulating specific lipid metabolic pathways. Etomoxir and AICAR have been shown previously to inhibit and up-regulate β-oxidation, respectively. While etomoxir did not rescue ROS or viability in palmitate-treated cells, AICAR limited ROS production and prevented cell death at concentrations between 250-500 μM. Phospholipase inhibitors have also been applied to modulate PC turnover. Phospholipase A_2 inhibitor $PACOCF_3$ demonstrated a dose-dependent reduction of ROS and caspase 3/7 activity. Preliminary results indicate that phospholipase C inhibitor U73122 can also successfully reduce ROS accumulation and rescue the 24-hour viability of H4IIEC3 cells.

Our ongoing work is focused on a detailed investigation of the mechanisms by which AICAR, PACOCF3 and U73122 are able to reverse the lipotoxic effects of SFA overexposure. We are applying lipidomic profiling and metabolic flux analysis (MFA) based on isotopic enrichment with ^{13}C-labeled palmitate to understand the metabolic fate of exogenous SFA and how it is altered by these interventions. This involves HPLC separation and GC-MS analysis of lipid extracts, followed by computational analysis of lipidomic and isotopic labeling data to identify pathways that are responsible for the observed metabolic phenotypes. Developing a better understanding of the mechanisms of SFA overexposure leading to ROS accumulation and lipoapotosis, and the identity of the specific lipid species mediating these events, will lead to more efficient and targeted therapies for treatment of fatty liver disease and related complications associated with diabetes and obesity.

See more of this Session: Proteomics & Metabolomic Approaches to Systems Biology

(182c) Directed Evolution of An E. Coli Methionyl-tRNA Synthetase for Two Strain Cell-Selective Proteomic Labeling

Frank Truong[1], Tae Hyeon Yoo[2], Thomas J. Lampo[1] and David A. Tirrell[1], (1)Division of Chemistry and Chemical Engineering, California Institute of Technology, Pasadena, CA, (2)Department of Molecular Science and Technology, Ajou University, Suwon, South Korea

Cell-selective metabolic labeling of proteins has enabled study of proteomic changes in a single subpopulation of a complex multi-cellular environment. Currently this method is limited to a single amino acid analog, azidonorleucine (Anl), using an engineered methionyl-tRNA synthetase (NLL-MetRS) for incorporation at codons designated for methionine (Met). To expand the set of tools for this method, we searched for other amino acid analogues capable of bio-orthogonal reactions and minimal activity with the endogenous *Escherichia coli* MetRS. Propargylglycine (Pra) was chosen as the target amino acid since the terminal alkyne side chain of Pra can undergo a [3+2] copper(I)-catalyzed cycloaddition with azides, allowing for conjugation of fluorescence probes and affinity tags to proteins bearing Pra residues in lieu of Met. Using both active-site randomization and error-prone PCR, two rounds of directed evolution yielded a MetRS variant capable of incorporating propargylglycine at near-quantitative levels (Pra-MetRS) in methionine-depleted media using a methionine-auxotrophic *E. coli* strain. Proteins synthesized in *E. coli* cells expressing Pra-MetRS could be detectably labeled with Pra in the presence of methionine in the growth medium as observed by in-gel fluorescence after click chemistry with Cy5-azide. Analysis with MALDI-TOF mass spectrometry revealed that replacement of methionine by Pra was approximately 10%. We sought to take advantage of NLL-MetRS and Pra-MetRS in order to cell-specifically label the proteomes of two bacterial strains with azidonorleucine and propargylglycine. An *E. coli* strain bearing a plasmid encoding an inducible His-tagged DHFR was co-expressed with the NLL-MetRS, while a second strain bearing a plasmid encoding an inducible His-tagged GFP variant was co-expressed with the Pra-MetRS. The two strains were co-cultured in media containing Anl, Pra, and Met and marker proteins were expressed and purified. Copper-free click chemistry with dibenzyl-aza-cyclooctyne-AlexaFluor488 was performed first to selectively label DHFR. Subsequently, copper(I)-catalyzed click chemistry with Cy5-azide resulted in selective labeling of GFP. Future work will focus on characterizing Pra-MetRS as well as using azidonorleucine and propargylglycine to label proteomes of similarly sized organisms to study a variety of interactions.

See more of this Session: Proteomics & Metabolomic Approaches to Systems Biology
See more of this Group/Topical: Topical A: Systems Biology

(182d) Inference of Predominant Pathways In Cancer: A Proteomic Study Among Three Human Mammary Epithelial Cell Models

Yogesh Kulkarni and **David Klinke**, Chemical Engineering, West Virginia University, Morgantown, WV

A contemporary view of the cancer genome reveals extensive rearrangement compared to normal cells. Yet how these genetic alterations translate into specific proteomic changes that underpin acquiring the hallmarks of cancer remains unresolved. The objective of this study was to quantify alterations in protein expression in three cellular models of breast cancer and infer differentially regulated signaling pathways associated with the hallmarks of cancer. A proteomic workflow was used to identify proteins in two HER2 positive tumorigenic cell lines (BT474 and SKBR3) that were differentially expressed relative to a normal human mammary epithelial cell line (184A1). Pathway inference tools were used to interpret these proteins in terms of functionally enriched pathways in the tumor cell lines. A total of 64 (BT474-184A1) and 69 (SKBR3-184A1) proteins were uniquely identified that were differentially expressed by at least 1.5-fold. We observed cell motility pathways to be predominantly associated with BT474 cell line, whereas amino acid metabolism pathways were associated with SKBR3 cell line. While "protein ubiquitination" and "apoptosis signaling" pathways were common to both the cell lines, the observed patterns of protein expression and post-translational modification suggest that the evasion of apoptosis in the tumorigenic cell lines occurs via different mechanisms.

See more of this Session: Proteomics & Metabolomic Approaches to Systems Biology
See more of this Group/Topical: Topical A: Systems Biology

(182e) Metabolomic Analysis of Ovarian Cancer

Kathleen Vermeersch and **Mark P. Styczynski**, School of Chemical & Biomolecular Engineering, Georgia Institute of Technology, Atlanta, GA

Here, we present a metabolomic analysis of two ovarian cancer cell lines: OVCAR-3, and a tumor-initiating cell line derived from OVCAR-3. We aim to characterize the metabolic differences between these two lines in physiologically relevant perturbations representative of conditions encountered in the tumor microenvironment. Conditions studied include carbon source deprivation, hypoxia, pH change, and chemotherapy treatment. Additionally, we also aim to characterize the metabolic phenotype associated with the mesenchymal-epithelial transition (MET).

The contribution of dysfunctional metabolism to the progression and phenotypes of cancer is increasingly being recognized, despite a decades-long stretch when genetics drove the vast majority of cancer research. While it is now well-known that cancers are usually characterized by an increase in glycolytic flux and fermentation, the extent of the impact of metabolism on cancer's phenotypes and progression is still uncertain. We hypothesize that there is a distinct metabolic signature for the two cell lines under study in different physiological conditions that contributes to their different phenotypes – most notably, the resistance to chemotherapy that is characteristic of tumor-initiating cells. Additionally, we also hypothesize that there are gross metabolic changes that characterize the mesenchymal-epithelial transition.

Here, we use two-dimensional gas chromatography coupled to mass spectrometry (GCxGC-MS) to measure intracellular and extracellular metabolites in ovarian cancer cell lines. As expected, we are able to capture metabolic changes in response to different perturbations, both on an individual metabolite level as well as via dimensional reduction analysis of the entire dataset. We report the differences observed between the two cell lines' metabolic responses and present a pathway-based analysis of the metabolic behaviors observed.

See more of this Session: Proteomics & Metabolomic Approaches to Systems Biology
See more of this Group/Topical: Topical A: Systems Biology

(182f) Effects of Small-Molecule Inhibition of Central Metabolism in a B-Cell Lymphoma Model of Myc Driven Cancer

Taylor A. Murphy and Jamey Young, Chemical and Biomolecular Engineering, Vanderbilt University, Nashville, TN

Cancer metabolism has, in recent years, become an attractive avenue for targeted therapeutics. Anabolic and catabolic processes are tightly regulated within any proliferating cell, with a cancerous cell being no different. Central metabolism supplies the cell with intermediate metabolites that satisfy energy, redox homeostasis, and biosynthetic requirements. The overall metabolic state is determined by the oncogenic mutations and microenvironmental conditions that characterize a particular cancer cell. The transcription factor c-Myc is an oncoprotein that exhibits deregulated expression in over 30% of all cancers. In addition to its role as a global regulator of cell cycle and apoptosis, c-Myc exerts significant control over cellular metabolism. Targeting deregulated metabolic pathways with small-molecule inhibitors has become an attractive therapeutic strategy because of recent findings that suggest these metabolic alterations may play a critical role in promoting cancer progression.

Our recent work using a B-cell line with regulatable expression of c-Myc reveals a strong coupling between c-Myc expression and metabolic state. We applied metabolic flux analysis (MFA) using ^{13}C-labeled glucose tracers and GC-MS profiling to compare High and Low Myc states. Growth was inhibited by 50% under the Low Myc state, while specific rates of glucose uptake and lactate secretion were decreased by 20% and uptake of several amino acids such as glutamine and serine were decreased by over 50% in comparison to the High Myc state. We have also treated High and Low Myc expressing cells with selective metabolic inhibitors to determine their effects on growth and metabolism. We hypothesized that the High Myc cells would be more sensitive to these inhibitors than the Low Myc cells. Early results indicate that High Myc cells are more sensitive to oxamate, which is a specific inhibitor of lactate dehydrogenase, with an IC50 an order of magnitude lower than that of the Low Myc cells. Ongoing work is being conducted to quantify the metabolic alterations that occur in response to this and other metabolic inhibitors using ^{13}C-MFA, and how they differ depending on the level of c-Myc expression.

See more of this Session: Proteomics & Metabolomic Approaches to Systems Biology
See more of this Group/Topical: Topical A: Systems Biology

(243a) Mechanochemical Energy Transduction In Muscle Contraction

Richard L. Long, New Mexico State University, Las Cruces, NM

Coupling of the linear momentum balance with the biochemical cycle and a model of force-velocity link in the crossbridge cycle enables computation of the steady state and transient dynamics of muscle action. Some controversies remain in this topic. for instance, some authors include lattice compliance while some data show that lattice compliance does not lead to very geat differences in simuation results at the macro scale. Anothe issue is the extent to which calcium ion transient are
required in modeling mechanical action versus using a rapid equilbrium model for these steps. The art of modeling these systems is to include enough complexity without introducing too much complexity.

See more of this Session: Multiscale Systems Biology
See more of this Group/Topical: Topical A: Systems Biology

(243b) Multiscale Variability In Human Endotoxemia: Circadian, Ultradian, and Higher Frequency Rhythms In Heart Rate Variability

Jeremy D. Scheff[1], Steven E. Calvano[2], Stephen F. Lowry[2] and Ioannis P. Androulakis[3], (1)Biomedical Engineering, Rutgers University, Piscataway, NJ, (2)Department of Surgery, UMDNJ-Robert Wood Johnson Medical School, New Brunswick, NJ, (3)Biomedical Engineering, Chemical & Biochemical Engineering, Rutgers University, Piscataway, NJ

Homeostasis exhibits rich variability, from the discrete activation of gene expression in individual cells to rhythms in hormone concentrations that regulate entire organs. Heart rate variability (HRV), the quantification of the distribution of time intervals between heartbeats, is one such variable physiological signal with particular clinical importance. Changes in HRV reflect alterations in autonomic function, which is potentially a useful predictor of outcome in myocardial infarction, congestive heart failure, diabetic neuropathy, and neonatal sepsis. Dysregulated autonomic function as assessed by HRV has been observed in intensive care units and has been hypothesized to have value as a marker of recovery from critical illness (Lowry and Calvano 2008). Due to this clinical relevance, dynamic characteristics of HRV have been assessed by time domain, frequency domain, and nonlinear metrics. The majority of HRV research has thus far focused on the interpretation of the patterns of HRV rather than linking cellular-level mechanisms to patterns. The realization that health may be characterized by a certain degree of variability of human heart signals motivates the hypothesis that appropriate physiologic variability is the manifestation of robust dynamics of control signals whose fluctuations equip the host with the ability to anticipate external and internal disturbances. We hypothesize that these variable dynamics are driven by the convergence of rhythmic physiological signals on the heart via autonomic modulation.

Studying the effects of critical illness on HRV requires a clinical model that can be experimentally evaluated in great detail. Human endotoxemia, the injection of lipopolysaccharides (LPS or endotoxin, used interchangeable herein) into healthy human subjects, has been extensively used as a model of systemic inflammation due to qualitatively similar responses in systemic physiologic and metabolic processes, including changes in leukocyte abundance and behavior, hormonal secretion, and cardiac function. Responses observed in human endotoxemia experiments mimic observed responses in systemic inflammation in ICU patients, albeit over different timescales, thus making the human endotoxemia model an excellent platform for exploring mechanistic underpinnings of the systemic inflammatory response. A key component in the response to endotoxemia is a decrease in HRV, concomitant with imbalances in autonomic activity reflected by perturbed autonomic oscillatory responses in HR.

Despite an understanding of the importance of inflammation in a wide variety of disorders and a large number of experiments elucidating the details of the inflammatory response, novel treatments aimed at controlling inflammation remain elusive. The complexity of the interacting, redundant pathways involved in the inflammatory response necessitate a systems-level understanding of inflammation, thus leading to interest in the inflammatory response from a systems biology perspective. The dynamic signals evoked in an inflammatory response are propagated to the sinoatrial (SA) node of the heart to assess how HRV is perturbed in endotoxemia. Previously, endotoxemia-induced changes in HR and HRV have been described by physicochemical relations which begin to elucidate the signals that give rise to altered phenotypes (Foteinou, Calvano et al. 2010). However, this neglects that HR and HRV are both derived from the same physiological process, the beats of the heart, and that the contraction of the heart as initiated by firing neurons at the SA node is a noisy, discrete process. This motivates the development of a more mechanistic model to produce discrete heartbeat signals that can then be used to calculate HR and HRV, providing a basis for the development of autonomic dysfunction in endotoxemia. We propose a semi-mechanistic mathematical model linking endotoxemia to cardiac function through an integral pulse frequency modulation (IPFM) model that produces discrete heartbeats as output based on autonomic modulation of the heart. Outputs of the model, namely HR and HRV, are shown to accurately capture experimentally-observed phenomena in human endotoxemia studies.

In this work, we further explore how physiologic variability outside the heart influences HRV. Therefore, based on the framework described above, we study how physiologic variability at multiple levels culminates in perturbing the host towards a dysregulated autonomic state in inflammation. Circadian rhythms, both in the suprachiasmatic nucleus (SCN) and in peripheral circadian oscillators, play a key role in the dynamics of inflammation. Many components of the inflammatory response are under circadian regulation (Scheff, Calvano et al. 2010), and it has been shown that endotoxemia suppresses the expression of clock genes that comprise the peripheral circadian oscillators in leukocytes (Haimovich, Calvano et al. 2010). The interplay between peripheral circadian rhythms and inflammatory cytokines perturbs the state of the host in an acute response to endotoxemia.

Ultradian rhythms in the secretion of cortisol, a key anti-inflammatory hormone, also play an important role in maintaining homeostasis systems that respond to glucocorticoids (Lightman and Conway-Campbell 2010). Glucocorticoid responsive genes are expressed at appropriate levels in the presence of ultradian rhythms, but when these rhythms are lost a large number of genes are dysregulated. As glucocorticoids are known to regulate the expression of key inflammatory genes, including pro-inflammatory and anti-inflammatory cytokines, and as pulsatility in hormone secretion may be lost in stress, the incorporation of ultradian rhythms into our integrated model alters the state of the inflammatory system as a whole in endotoxemia, which is reflected in HRV.

Our model is motivated and validated by data at multiple scales. The human endotoxemia response (Lowry 2005) is well characterized from the perspective of hormonal and cytokine measurements. Gene expression data is also available, which we leverage to model the evolving transcriptional response to endotoxemia. To study model responses under a variety of experimental conditions, we use data from experiments where hormone treatment is given prior to endotoxin. Furthermore, model subsystems, such as the ultradian release of glucocorticoids by the HPA axis, can be studied in isolation based on the wide range of experimental data on glucocorticoid release and action.

Understanding the loss of variability of cardiac function in endotoxemia serves as a step towards gaining insight into similar changes in HRV observed clinically in response to stress. This work builds towards translational applications of systems biology by moving towards an understanding of the relationship between fundamental biological processes and clinical outcomes.

References

- Foteinou, P. T., S. E. Calvano, et al. (2010). "Multiscale model for the assessment of autonomic dysfunction in human endotoxemia." Physiol Genomics 42(1): 5-19.
- Haimovich, B., J. Calvano, et al. (2010). "In vivo endotoxin synchronizes and suppresses clock gene expression in human peripheral blood leukocytes." Crit Care Med 38(3): 751-8.
- Lightman, S. L. and B. L. Conway-Campbell (2010). "The crucial role of pulsatile activity of the HPA axis for continuous dynamic equilibration." Nat Rev Neurosci 11(10): 710-8.
- Lowry, S. F. (2005). "Human endotoxemia: a model for mechanistic insight and therapeutic targeting." Shock 24 Suppl 1: 94-100.
- Lowry, S. F. and S. E. Calvano (2008). "Challenges for modeling and interpreting the complex biology of severe injury and inflammation." J Leukoc Biol 83(3): 553-7.
- Scheff, J. D., S. E. Calvano, et al. (2010). "Modeling the influence of circadian rhythms on the acute inflammatory response." J Theor Biol 264(3): 1068-76.

See more of this Session: Multiscale Systems Biology
See more of this Group/Topical: Topical A: Systems Biology

(243c) Patient-Specific Predictions of Thrombosis

Matthew H. Flamm, Tom Colace, Manash Chatterjee, Talid Sinno and Scott L Diamond, Chemical and Biomolecular Engineering, University of Pennsylvania, Philadelphia, PA

Plateletaggregation is an important regulator of hemostaticfunction in response to a blood vessel injury.Overactivity of platelet aggregation insituations such as atherosclerotic plaque rupture can cause occlusion of thevessel, while underactivity results in excessivebleeding. Several complex biological andphysical mechanisms contribute to the growth of a stable platelet mass at thesite of injury while limiting the final extent of growth to prevent occlusion. A detailed model of this system would provideinsight on the coupling between biological signaling and fluid flow as well aspredict strategies for anti- or pro-thrombotic therapy. Patient-specific models would enablepredictions of resistances to common anti-platelet therapies such as asipirin and clopidogrel. A multiscale modelis built upon patient-specific platelet phenotypingand is compared to patient-specific platelet deposition to collagen in a microfluidic chamber.

Themultiscale model has four major modules: latticeBoltzmann (LB), finite element (FEM), lattice kinetic Monte Carlo (LKMC), andneural network (NN). The NN is trainedon a single donor's pairwise agonist scanning (PAS)experiment that measures the rise in intracellular calcium to three mainplatelet agonists: adenosine diphosphate (ADP), thromboxane A2 (TXA2), and convulxin(CVX). The LKMC method follows themotion of platelets within the fluid due to diffusion and convection. Platelet bonding is captured in LKMC througha bonding model that links prediction of intracellular calcium from the NN tobonding kinetics. The concentrations ofsoluble platelet agonists, which are released from activated platelets, aretracked using FEM. Finally, LB solvesfor the velocity of the fluid around the growing platelet aggregate. Due to the efficiency of the multiscale framework, the entire active zone' of theexperiment can be simulated in 2 dimensions.

Themultiscale model is directly compared to experimentalresults of platelet aggregation for 3 donors with 3 antiplatelettherapies: COX inhibition, P_2Y_1 inhibition, and prostacyclin receptor activation. The model predicts that donor 1 hasconsistently larger platelet aggregates than donors 2 or 3, which is confirmedby experiment. The model also correctlypredicts the donor-specific potency of antiplatelettherapies. In general, prostacyclin receptor activation had the largest effect onplatelet aggregation due to a decrease in collagen signaling. P_2Y_1 inhibition andCOX inhibition both reduce propagation of the aggregate once a monolayer is formed, however P_2Y_1 inhibition has agreater effect. These trends are alsoseen in the experimental results. Themodel also tracks the morphology of the platelet aggregate and shear rate(force) distribution along the solid-fluid boundary.

.

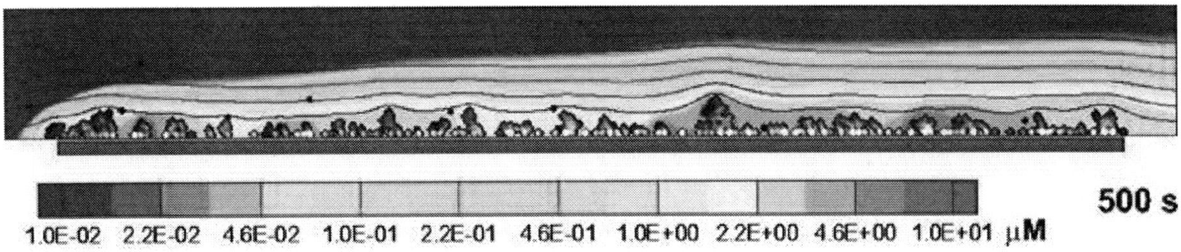

Figure 1. Simulation of platelet aggregation on collagen surface. Circles platelets (Black fully unactivated, White fully activated). Lines streamlines of the blood flow. Background color ADPconcentration (TxA_2 concentration not shown). Red bar collagen patch.

See more of this Session: Multiscale Systems Biology
See more of this Group/Topical: Topical A: Systems Biology

(243d) A Computational Model of Dynamic Flow In the Respiratory System

Aleck Alexopoulos, Chemical Process Engineering Research Institute, Thessaloniki, Greece and **Costas Kiparissides**, Department of Chemical Engineering, Aristotle University of Thessaloniki & Chemical Process Engineering Research Institute, Thessaloniki, Greece

Airflow in the pulmonary tract is important in terms of proper function of the respiratory system and adequate gas exchanger at the alveoli. In diseased states the inspired volume of air decreases and, in extreme cases, can lead to decreased in oxygen levels in the blood stream. Pulmonary airflow is also important in terms of particle transfer, deposition, and removal or penetration. The transfer and deposition of particles in the respiratory system is of major interest for the development of targeted local and systemic drug delivery formulations (Illum, 2006; Dalby and Suman, 2003, Azarmi et al., 2008). The effect of size and inhalation rate on particle transfer and deposition has been well studied for smooth spherical particles and steady inhalation. However, several issues remain to be elucidated including the deposition of non-spherical particles, dispersion and aggregation effects, the effect of particle charge and dynamic effects during a breathing cycle.

Deposited particles can undergo disaggregation, can release beneficial drugs, and, if sufficiently small, can penetrate into the bloodstream. The effective transfer and deposition of particles depends on the physiological state of the respiratory system, e.g. diseased or normal. Diseased physiological states can present constricted airflows, increased mucous layer thickness (due to overproduction or decreased clearance), and acute and chronic inflammation. The effective and targeted delivery of drugs in such diseased states remains a challenge to this day.

The pulmonary system consists of approximately 24 generations of airway branches for a total of about $3\ 10^7$ branches. The dynamics of flow change dramatically from the bronchi to the alveoli as airway diameters decrease from 6mm to 100μ m, velocities from 5m/s to 2.5mm/s, and local Reynolds numbers from about 1000 to 0.1. Consequently a full comprehensive (e.g., CFD) computational model of flow in the pulmonary system let alone a full realistic geometric representation is out of the question (Finlay, 2004). Consequently, approximate integrated models are being pursued. Integrated multi-scale physiological models of the respiratory system have been under development over the past few years (Burrowes et al., 2008). Integrative models often describe a small part of the pulmonary system (Martonen et al., 2002; Zhang et al., 2002) or employed sequences of branches or groups of branches to describe the flow from the bronchi to the alveoli (Nowak et al., 2003; Zhang et al., 2009). The necessity of an accurate geometric description of the URT has also become evident in recent simulations (van Ertbruggen et al.; 2007, Longest and Vinchurkar; 2007). Despite this progress, no computational models integrating the respiratory system from the bronchi to the alveoli nor any realistic models for dynamic flow have been reported.

To further improve our understanding of airflow as well as particle deposition during a breathing cycle a multi-level computational model of the respiratory system was developed that describes the dynamic airflow and motion of particles through the lungs. In this work airflow through the pulmonary system down to the alveolar sac and individual alveoli level is determined by dynamic and steady-state CFD simulations. The pulmonary system is separated into the upper respiratory tract, URT, consisting of 5-8 generations, the lower respiratory tract, LRT, consisting of multiple branching structures (bronchioles) and the alveolar sacs.

The URT is described by either a symmetrical branched geometry or a physiologically-realistic geometry. Airflow in the URT is described by a transitional SST k-ω turbulence model while particle depositions are described by a Lagrangian-particle/Eulerian-fluid approach. The lower respiratory tract, LRT, is described by a simplified laminar-flow model, with averaged resistances to flow. Deposition of inertial particles in the LRT is assumed to follow a dimensionless universal law dependent only on the local Stokes number (Zhang et al., 2009).

The alveolar sac model consists of spherical caps inflating and deflating instantaneously during a breathing cycle. According to this model, during inhalation the initial abrupt decrease in alveoli pressure drives the airflow towards the alveoli. During the inhalation stage inflowing air to the alveoli gradually increase the pressure reducing the driving force for flow. At the end of the inhalation period, the alveoli volume changes abruptly and the sudden increase in alveoli pressure drives exhalation. The transfer of pressure and airflow rates through the branched structure (i.e., from the alveoli to the URT) is described with a simple quasi-steady state laminar-flow model.

This paper reports on the effects of dynamic flow on airflow as well as the transfer and deposition of particles throughout the entire respiratory system for different inhalation rates and different sized inertial particles. The effects of pathophysiological conditions on resistance, airflow, and particle deposition is also shown.

References

Azarmi, S., W.H. Roa, R. Lobenberg, 2008, Targeted delivery of nanoparticles for the treatment of lung diseases. Advanced Drug Delivery Reviews, 60, 863-875.

Burrowes, K.S. A.J. Swan, N.J. Warren, and M.H. Tawhai, 2008, Towards a virtual lung: multi-scale, multi-physics modeling of the pulmonary system, Phil. Trans. R. Soc. A, 366, 3247-3263.

Dalby R. and J. Suman, 2003, Inhalation therapy: technological milestones in asthma treatment. Advanced Drug Delivery Reviews, 55, 779-791.

Finlay, W.H. "Lung Deposition Simulation" in A.J. Hickey (ed.) Pharmaceutical Inhalation Aerosol Technology, Marcel Dekker, NY 2004.

Illum, L., 2006, Nanoparticulate systems for nasal delivery of drugs: a real improvement over simple systems? J. Pharm. Sci. 96, 473-483.

Martonen, T.B. Z. Zhang, G. Yue, and C.J. Musante, 2002, "3-D particle transport within the human upper respiratory tract. Aerosol Science 33, 1095-1110.

Longest, P. and Vinchurkar, S., 2007, "Validating CFD predictions of respiratory aerosol deposition: Effects of upstream transition and turbulence", J. of Biomechanics, 40(20, 305-316

Nowak N., P.P. Kakade, and A.V. Annapragada, 2003, "Computational Fluid Dynamics Simulation of Airflow and Aerosol Deposition in Human Lungs", Annals of Biomed. Engineering, 31, 374-390.

van Ertbruggen, C., C. Hirsch, and M. Paiva, 2005, "Anatomically based three-dimensional model of airways to simulate flow and particle transport using computational fluid dynamics", J. Appl. Physiol. 98, 970-980.

Zhang Z., C. Kleinstreuer, and S.K. Chong, 2002, "Aerosol deposition efficiencies and upstream release positions for different inhalation modes in an upper bronchial airway model" Aerosol Sci. & Technol, 36, 828-844.

Zhang, Z. C. Kleinstreuer, and C.S. Kim, 2009, "Comparison of analytical and CFD models with regard to micron particle deposition in a human 16-generation tracheobronchial airway model" Aerosol Science, 40, 16-28.

See more of this Session: Multiscale Systems Biology
See more of this Group/Topical: Topical A: Systems Biology

(243e) Nonlinear Dynamics of Poly(β-hydroxybutyrate) Production In Microorganisms

Andr Franz[1], Hartmut Grammel[2], Ruxandra Rehner[2] and Achim Kienle[3], (1)Process Synthesis and Process Dynamics, Max-Planck-Institute for Dynamics of Complex Technical Systems, Magdeburg, Germany, (2)Systems Biology, Max-Planck-Institute for Dynamics of Complex Technical Systems, Magdeburg, Germany, (3)Chair of Automation / Modeling, University of Magdeburg, Magdeburg, Germany

Poly(β-hydroxybutyrate) (PHB) is an organic polymer, which can be synthesized by many microorganisms under unbalanced growth conditions and which serves as internal energy and carbon reserve material. This bacterial polyester has useful properties for applications as thermoplastics, elastomers and adhesives and is biodegradable and biocompatible.

In the present paper focus is on nonlinear dynamics during PHB formation triggered by the underlying regulatory processes. Depending on the microorganism different patterns of behavior are observed. Focus is on *Ralstonia eutropha* (also known as *Alcaligenes eutrophus* and *Cupriavidus necator*) compared to *Rhodosprillum rubrum.*

PHB production in *Ralstonia eutropha* has been studied intensively in the past. Recently, the existence of multiple steady states and bistability in continuous cultures of *R. eutropha* has been discussed in the literature based on a simplified metabolic network leading to a simple cybernetic model [1], which assumes optimal regulation in view of limited resources. In the present paper, it is shown that the multiplicity/bistability region tends to vanish if a more detailed model is used, which is fitted to experimental data for PHB production and PHB consumption [2]. Our model is based on the hybrid cybernetic modeling approach which provides a systematic framework for dynamic model generation from metabolic flux analysis. It shows very good agreement with independent experimental data over a wide range of operating conditions.

Different patterns of behavior are observed for *Rhodospirillum rubrum*. Depending on the ambient conditions *R. rubrum* is able to switch between respiratory and photosynthetic metabolism and thereby offers new perspectives for biotechnological production processes. Experimental data from microscopy and flow cytometry during growth on acetate reveal transient bimodal population dynamics in a batch reactor indicating bistability on the single cell level. An initially homogeneous population is segregating into a first subpopulation with only little amount of PHB and a second subpopulation with relatively high amount of PHB. Both subpopulations are merging again at later time points to build a homogeneous population again. Potential sources on the single cell level are discussed using metabolic flux analysis. Quantitative modeling approaches on the single cell level and their integration into the population level will be presented.

References:

M. A. Pinto, C. D. Immanuel, Sensitivity of bifurcation traits to model parameters in poly-beta-hydroxy butyrate production, In: ADCHEM 2006

A. Franz, H.-S. Song, D. Ramkrishna, A. Kienle, Experimental and theoretical analysis of poly(β-hydroxybutyrate) formation and consumption in *Ralstonia eutropha*, Bio. Eng. J. **55** (2011) pp. 49-58

See more of this Session: Multiscale Systems Biology
See more of this Group/Topical: Topical A: Systems Biology

(243f) Stochastic Operator Splitting Method for Biological Systems

TaiJung Choi[1], **Mano R. Maurya**[2], Daniel M. Tartakovsky[1] and Shankar Subramaniam[3], (1) Department of Mechanical and Aerospace Engineering, University of California, San Diego, La Jolla, CA, (2)Department of Bioengineering, University of California San Diego, La Jolla, CA, (3)Department of Bioengineering, San Diego Supercomputer Center, Department of Chemistry & Biochemistry, University of California San Diego, La Jolla, CA

Stochastic OperatorSplitting Method for Biological SystemsTaiJungChoia,Mano Ram Maury[ab,a], andShankar Subramaniam[b,c,d,1]

[a]*Departmentof Mechanical and Aerospace Engineering* [b] *Departmentof Bioengineering* [c]*GraduateProgram in Bioinformatics* [d]*Departmentof Chemistry & BiochemistryUniversity of California, SanDiego, 9500 Gilman Dr La Jolla, CA 92093*E-mail addresses:TaiJungChoi: **tjchoi@ucsd.edu**ManoRam Maurya: **mano@sdsc.edu**DanielM. Tartakovsky: **dmt@ucsd.edu**ShankarSubramaniam: **Shankar@ucsd.edu**Deterministic modelsof biological and biochemical processes at the sub-cellular levelmight become improper when a series of chemical reactions is executedby a few molecules. Inherent randomness of such phenomena calls forthe use of stochastic simulations. Moreover, in case of inhomogeneousenvironment, such as receptor induced signaling in membrane andchemotaxis due to existence of chemoattractants , spatial effect alsoshould be considered. Therefore, diffusion phenomenon induced bychemical gradient becomes another important factor in biologicalanalysis. However, being computationally intensive, such simulationsbecome infeasible for large and complex reaction networks. To improvetheir computational efficiency in handling these networks, we presentan operator splitting approach(1), in which reactions are handledthrough exact stochastic simulation like Gillespie algorithm (2) anddiffusions are treated through Brownian dynamics. The proposedoperator splitting algorithm is used to model the simplified DNA/RNAsynthesis and reactions/diffusions of CheY molecules through thecytoplasm of Escherichia coli(3). At relatively high concentrations,the response characteristics obtained with the stochastic anddeterministic simulations coincide, which validates both approaches.At low doses, the response characteristics of some key chemicalspecies, such as levels of CheY, predicted with stochasticsimulations differ quantitatively from their deterministiccounterparts.

References:

RodrguezVidal J, et al. Spatial stochastic modelling of thephosphoenolpyruvate-dependent phosphotransferase (PTS) pathway inEscherichia coli. Bioinformatics 2006;22:1895-1901.Gillespie,D. T. 1976. General method for numerically simulating stochastic timeevolution of coupled chemical-reactions. Journal of ComputationalPhysics. 22:403-434.LipkowK, et al.Simulated diffusion of phosphorylated CheY through the cytoplasm ofEscherichia coli. J. Bacteriol. 2005;187:45-53.

1 Corresponding author: E-mail: shankar@sdsc.edu

See more of this Session: Multiscale Systems Biology
See more of this Group/Topical: Topical A: Systems Biology

(313a) How Conjugation Contributes to Antibiotic Resistance of Biofilm - from an Aspect of Population Balance Model with Stochastic Intracellular Gene Regulation

Che-Chi Shu[1], Doraiswami Ramkrishna[1], Anushree Chatterjee[2] and Wei-Shou Hu[2], (1)Chemical Engineering, Purdue University, West Lafayette, IN, (2)Chemical Engineering and Materials Science, University of Minnesota, Minneapolis, MN

In this work, **a** population balance model with stochastic intracellular generegulation is applied to analyze the important role of conjugation on biofilmdrug resistance. The transfer of conjugative plasmid helps bacteria spreadantibiotic resistance. Specifically, we focus on the conjugation of plasmidpCF10, which is influenced by both pheromone, secreted by recipients, andinhibitor, secreted by donors. Mass transfer effects in the biofilm alter theratio of pheromone to inhibitor of a cell, which in turn influlencesthe DNA configuration. Accordingly, the gene expression pattern is altered. Ourmodel predicts that biofilm has a pronounced effect on the induction ofconjugation. Further, the planktonicenvironment shows unimodal distributions around the on or off states, while thebiofilm environment shows bimodal distributions. While the addition of inhibitor has only asmall effect on the biofilm, the effect on cells in planktonic growth issubstantial. These results offer a plausible explanation of biofilm antibioticresistance caused by conjugation.

See more of this Session: Modeling Approaches In the Life Sciences I
See more of this Group/Topical: Topical A: Systems Biology

(313b) Modeling Circadian Interactions Between Peripheral Clocks and Metabolism

Panagiota T. Foteinou and Francis J. Doyle III, Chemical Engineering, University of California, Santa Barbara, CA

The circadian system or "biological clock" orchestrates the temporal organization of several essential physiological processes. This system enables the organism to adapt to common daily changes, such as day-night cycle and food availability. In mammals, the central pacemaker resides within the suprachiasmatic nucleus (SCN), which receives entraining signals from the environment and coordinates the oscillating activity of peripheral clocks that are present in almost all tissues. Circadian oscillators share the same molecular make up in SCN neurons and peripheral cell types. The predominant external cue of the central clock is light which, at the level of SCN neurons, stimulates a cascade of signaling pathways that lead to the activation of a transcriptional program. Peripheral tissues also contain functional circadian oscillators that are self-sustained at the single cell level, but they do not respond to light-dark cycles. Instead, cycling feeding behavior is a strong timing cue for peripheral clocks which are important for driving local rhythmic activity, probably through modifications of nutrient availability. While many metabolic functions are subjected to circadian variation metabolic signals also feedback into the circadian system (Green, Takahashi et al. 2008). Such observation reflects the now recognized relationship between circadian transcription and metabolic regulatory networks.

Disruption of circadian rhythms has a profound impact on several pathologies including mental, behavioral and metabolic disorders (Bass and Takahashi). Linking circadian misalignment to physiological behavior requires a systems-level understanding of the hierarchical organization of the biological clock. A vital enabler in that respect is the development of in silico methodologies that integrate circadian regulatory processes at various scales (brain, periphery). Prior modeling efforts from our group have focused on the development of a cell autonomous mathematical model of a circadian neuron (Mirsky, Liu et al. 2009). Due to the critical role peripheral clocks play in health and disease, the work to be discussed here extends such an integrative approach in assessing the impact of circadian dysregulation in peripheral tissue pathophysiology. Using quantitative approaches to reverse-engineer the mechanisms underlying the generation of circadian rhythms, we expand the envelope of circadian interactions by quantifying the regulation of cell autonomous circadian rhythms by metabolic sensors.

Of particular relevance to our study are a few recent groundbreaking studies (Asher, Gatfield et al. 2008; Nakahata, Kaluzova et al. 2008) showing that the NAD^+-dependent enzyme SIRT1 functions as a histone deacetylase that counteracts the histone acetyltransferase activity (HAT) of CLOCK. Both studies show that the histone deacetylase SIRT1 is involved in regulating the amplitude of circadian clock-controlled gene expression. Further studies (Nakahata, Sahar et al. 2009; Ramsey, Yoshino et al. 2009) went on by demonstrating that, in turn, the clock regulates the expression of nicotinamide phosphoribosyltransferase the rate limiting enzyme in the NAD^+ salvage pathway, as it catalyzes the first step in the biosynthesis of NAD^+. These findings imply that not only the clock modulates metabolism but also the other way around, the metabolic status of the cell influences the clock machinery; thus generating an interlocking of the transcriptional feedback clock loop with the enzymatic loop of the NAD^+-salvage pathway (Eckel-Mahan and Sassone-Corsi 2009). It is therefore the purpose of this research to quantify such reciprocal regulation of clock making it a critical enabler for assessing the impact of metabolic activity on the circadian clock machinery.

To meet this challenge, a cell autonomous model that quantifies the reciprocal regulation of clock by metabolic signaling has been developed. This work operates along the following three axes: (i) generation of circadian (self-sustained) oscillations governed by the canonical PER-CRY/CLOCK-BMAL1 feedback loop, (ii) circadian control of the enzymatic NAD^+-salvage pathway and finally (iii) integration of the mammalian circadian clock with intracellular metabolism through the oscillating biosynthesis of NAD^+. This new time-keeping loop is closed by feedback of SIRT1 on CLOCK-BMAL1 dependent transcription by counteracting CLOCK HAT activity, particularly inhibiting CLOCK-mediated BMAL1 acetylation. Since CLOCK-dependent BMAL1 acetylation potentiates its binding by PER-CRY repressor (Hirayama, Sahar et al. 2007), we explore the hypothesis that SIRT1 inhibits such interaction. Driven by the premise that SIRT1 mRNA accumulation does not display robust circadian rhythm (similar to the almost constant levels of CLOCK expression in most mammalian tissues), SIRT1 mRNA dynamics are not explicitly modeled. Instead, its enzymatic activity oscillates in phase with NAD^+ levels and therefore the dynamics of NAD^+ metabolite serve as a surrogate signal for SIRT1 deacetylase activity.

The performance of the proposed model is evaluated through its potential to capture experimentally observed cell-autonomous phenotypes of gene knockouts including circadian amplitude variation due to loss of metabolic activity (e.g. $SIRT1^{-/-}$). Interestingly, the simulated increase in the amplitude of oscillations due to lack of SIRT1 is consistent with the experimental findings as reported by Nakahata et al. (Nakahata, Kaluzova et al. 2008). In another study (Asher, Gatfield et al. 2008), Asher et al. propose another mechanism by which SIRT1 activity might impinge on the circadian rhythm, possibly through SIRT1-mediated PER deacetylation. By considering that such deacetylation occurs while PER protein interacts with CLOCK-BMAL complex facilitating its degradation rate, the simulated circadian amplitude decreases (rather than increases) under SIRT1 knockout conditions. Such simulations capture the experimentally observed differential response of clock gene expression to loss of SIRT1. Further modeling results involve simulations related to circadian dysfunction under mutations of clock genes (e. g. $BMAL1^{-/-}$, $CRY^{-/-}$). In addition to arrhythmias, the model simulates lower constitutive levels of clock-controlled genes due to loss of the essential BMAL1 activator while greater expression levels are simulated for the CRY double knockout phenotype.

This work demonstrates the feasibility of a circadian transcriptional-enzymatic model that captures for the first time metabolic and physiological core components of the overall control of the mammalian circadian rhythm. Modeling the reciprocal regulation of clock by metabolic signaling provides the necessary ammunition to move forward with the proposed model assessing the influence of circadian rhythms on metabolic homeostasis. By integrating how metabolism impinges upon the circadian rhythm and vice versa, the ultimate goal is to explore mechanisms of synchronization between central and peripheral clocks both at the single cell and tissue level. Such *in silico* methodologies have the potential of elucidating putative mechanisms at various scales that can be functionally modulated for resetting the clock.

References

Asher, G., D. Gatfield, et al. (2008). "SIRT1 regulates circadian clock gene expression through PER2 deacetylation." Cell **134**(2): 317-28.

Bass, J. and J. S. Takahashi "Circadian integration of metabolism and energetics." Science **330**(6009): 1349-54.

Eckel-Mahan, K. and P. Sassone-Corsi (2009). "Metabolism control by the circadian clock and vice versa." Nat Struct Mol Biol **16**(5): 462-7.

Green, C. B., J. S. Takahashi, et al. (2008). "The meter of metabolism." Cell **134**(5): 728-42.

Hirayama, J., S. Sahar, et al. (2007). "CLOCK-mediated acetylation of BMAL1 controls circadian function." Nature **450**(7172): 1086-90.

Mirsky, H. P., A. C. Liu, et al. (2009). "A model of the cell-autonomous mammalian circadian clock." Proc Natl Acad Sci U S A **106**(27): 11107-12.

Nakahata, Y., M. Kaluzova, et al. (2008). "The NAD+-dependent deacetylase SIRT1 modulates CLOCK-mediated chromatin remodeling and circadian control." Cell **134**(2): 329-40.

Nakahata, Y., S. Sahar, et al. (2009). "Circadian Control of the NAD(+) Salvage Pathway by CLOCK-SIRT1." Science **324**(5927): 654-657.

Ramsey, K. M., J. Yoshino, et al. (2009). "Circadian clock feedback cycle through NAMPT-mediated NAD+ biosynthesis." Science **324**(5927): 651-4.

See more of this Session: Modeling Approaches In the Life Sciences I
See more of this Group/Topical: Topical A: Systems Biology

(313c) Metabolite-Centric Approach for Drug Targeting and Discovery Against Microbial Pathogens

Hyun Uk Kim, Tae Yong Kim and Sang Yup Lee, Korea Advanced Institute of Science and Technology (KAIST), Daejeon, South Korea

Extensive genomic knowledge on microbial pathogens has become available from a-decade-long efforts, yet there still exists a gap to fill in between genomics and actual drug discovery. Prompted by this biomedical issue, we developed a metabolite-centric approach for effective drug targeting and discovery against microbial pathogens, and, in particular, in this study, this approach was applied to an opportunistic human pathogen *Vibrio vulnificus* CMCP6 as a demonstration. This drug targeting method aims at predicting so called essential metabolites whose absence disrupts cell growth using a genome-scale metabolic network model of the target organism. For this, we first reconstructed the genome-scale metabolic network of *V. vulnificus* CMCP6, named VvuMBEL943, using its newly updated genomic information. Then, essential metabolites were predicted by implementing constraints-based flux analysis, which is an optimization-based simulation technique. This initial set of the predicted essential metabolites was screened with additional criteria of organism specificity and maximal disruptive damage to the cell. Essentiality of final five essential metabolites survived from the filtering was experimentally validated by gene knockout experiments. Finally, structural analogs of the final five essential metabolites were selected out of large chemical compound library, and employed for whole-cell screening to identify the most effective antibacterial candidate compound. This systematic approach should facilitate initial stage of drug discovery for antibiotics as well as other human diseases.

[This work was supported by the Korean Systems Biology Research Project (20100002164) and World Class University program (R322009000101420) of the Ministry of Education, Science and Technology through the National Research Foundation of Korea.]

See more of this Session: Modeling Approaches In the Life Sciences I
See more of this Group/Topical: Topical A: Systems Biology

(313d) Heterogeneous Microbial Populations: Using Flow Cytometric Data for Building Dynamic Distributed Models

Rita Lencastre Fernandes[1], Magnus Carlquist[2], Luisa Lundin[3], Anna-Lena Heins[4], Abhishek Dutta[5], Ingmar Nopens[5], Anker D. Jensen[1], Sren J. Johansen[3], Anna Eliasson Lantz[4] and Krist V. Gernaey[1], (1) Department of Chemical and Biochemical Engineering, Technical University of Denmark, Kgs. Lyngby, Denmark, (2)Department of Chemistry, Lund University, Lund, Sweden, (3)Department of Biology, University of Copenhagen, Copenhagen, Denmark, (4)Department of Systems Biology, Technical University of Denmark, Kgs. Lyngby, Denmark, (5)Biomath, University of Ghent, Ghent, Belgium

Traditionally, microbial populationshave been considered homogeneous in studies of fermentation processes. However,research has shown that a typical microbial population in a fermentor isheterogeneous [1-3]. Phenotypic heterogeneity arises as a result of the variability inherentto the metabolic processes within single cells. Two dominant cell variablesresponsible for differential gene expression are cell cycle and cell ageing [4].Indeed, cells at different phases in the cell cycle, or with different ages,have been observed to respond differently to stress conditions [1].

Although the number of experimental methods available for single-cellanalysis has boomed [5, 6], the knowledge acquired by such experimental studieshas not yet been integrated into a generally accepted modeling framework ableto account for distributed properties within a cell population [3].

In this work, focus was set on experimentally studying, as well asmodeling, the dynamics of phenotypic heterogeneous populations of *Saccharomyces cerevisiae*during batch cultivations. Besides the common monitored variables (e.g. opticaldensity, glucose, ethanol), single-cell total protein content and DNA contentwere measured by flow cytometry during the differentphases of batch cultivations. Aiming at establishing a population balance model(PBM) which describes the dynamic behavior of the yeastpopulation (including the relative contribution of different subpopulations), asystematic analysis of the flow cytometric data wasperformed, and mathematical descriptions for the budding initiation and celldivision rates as functions of the available substrate concentration areproposed.

[1] AverySV. Microbial cell individuality and the underlying sources of heterogeneity.Nat Rev Microbiol 2006; 4:577-587.

[2] Enfors SO, Jahic M, Rozkov A, Xu B, Hecker M, Jrgen B *et al.* Physiologicalresponses to mixing in large scale bioreactors. J Biotechnol 2001; 85:175-185.

[3] Mller S, Harms H, BleyT. Origin and analysis of microbial population heterogeneity in bioprocesses. Curr Opin Biotechnol 2010; 21:100-113.

[4] Sumner ER, Avery SV.Phenotypic heterogeneity: differential stress resistance among individual cellsof the yeast *Saccharomyces cerevisiae*.Microbiology 2002M; 148:345-351.

[5] Schmid A, Kortmann H,Dittrich PS, Blank LM. Chemical and biological single cell analysis. Curr OpinBiotechnol 2010; 21:12-20.

[6] LencastreFernandes R, Nierychlo M, Lundin L, Pedersen AE,Puentes Tellez PE, Dutta A *et al.* Experimentalmethods and modeling techniques for description of cell populationheterogeneity. Biotechnol Adv2011; doi:10.1016/j.biotechadv.2011.03.007

See more of this Session: Modeling Approaches In the Life Sciences I
See more of this Group/Topical: Topical A: Systems Biology

(313e) Efficient Simulation of Dynamic Flux Balance Models

Stuart Harwood, Kai Hoeffner and Paul Barton, Process Systems Engineering Laboratory, Dept. of Chemical Engineering, Massachusetts Institute of Technology, Cambridge, MA

Dynamic flux balance modeling is a powerful tool for the analysis and optimization of biochemical reactors. Dynamic flux balance analysis (DFBA) assumes that species within a cell rapidly reach equilibrium with the extracellular environment. Intracellular metabolism isthen coupled with the dynamic mass balance equations for the extracellularenvironment via expressions for the rates of substrate uptake and productexcretion. The model for the intracellular metabolism is a stoichiometric modelthat relates species and fluxes via a reaction network. However, this linearsystem is underdetermined since there are more fluxes than species. This isovercome by assuming that the cell will maximize some objective within theimposed flux constraints. If this objective, usually cell growth, can bemodeled as a linear function, the result is a linear program (LP). The overalldynamic flux balance model is an ordinary differential equation (ODE) with a LP"embedded".

The general form is

$\mathbf{x}'(t) = \mathbf{f}(t,\mathbf{x}(t),\mathbf{u}(t,\mathbf{x}(t))),\ \mathbf{x}(t_0) = \mathbf{x}_0$

with $\{\mathbf{u}(t,\mathbf{z})\} = \arg\min_{\mathbf{v}} \mathbf{c}^{\mathrm{T}}\mathbf{v}$ (1)

s.t. $\mathbf{A}\mathbf{v} = \mathbf{b}(t,\mathbf{z})$,

$\mathbf{v} \geq \mathbf{0}$.

These models have proven useful and accurate, andone could use these models to perform dynamic optimization of bioreactorsystems. However, solving (1) by embedding a LP solver in a code fornumerically integrating ODEs leads to quite an inefficient and inaccurateimplementation, especially if the dynamics are stiff, the LP is large and/orseveral LPs modeling several micro-organisms are embedded. The issue at hand isthat the vector field, which depends on the optimal solution to the LP, isnonsmooth. In addition, solving a LP is normally a more computationally complextask than a simple function evaluation and this induces numerical"noise". On the other hand, the algorithm described in this workovercomes both of these issues, and the result is an efficient solutionprocedure for problems of the form (1).

A code implementing the proposed algorithm has been written,called DSL48LPR. In the most general sense, DSL48LPR solves the LP once usingthe simplex algorithm and records the basis set B, an index set which partitions theprimal variables $\mathbf{v}$. The solution set (whilethis basis is optimal) is obtained very simply. The basic variables $\mathbf{v}_B$ of thesolution set are given by the solution to $\mathbf{A}_B\mathbf{v}_B = \mathbf{b}(t,\mathbf{z})$(where $\mathbf{A}_B$ is asubmatrix constructed from the columns of $\mathbf{A}$corresponding to B)and the nonbasic variables are simply zero. For the LP in (1) a basis set isoptimal until it becomes infeasible; since the LP is in standard from, this isindicated by a variable crossing zero. Using the event detection capabilitiesof DAEPACK component DSL48E, integration stops when a variable crosses zero.The basis set is updated by re-solving the LP before continuing integration.Thus, for most integration steps, a system of linear equations is solved,rather than a linear program. Overall, (1) is reformulated asdifferential-algebraic equations (DAEs) with discrete modes. This frameworkprovides an efficient and accurate method for solving a system of ODEs withembedded LPs.

See more of this Session: Modeling Approaches In the Life Sciences I
See more of this Group/Topical: Topical A: Systems Biology

(313f) An In-Silico Evolution Algorithm for Obtaining Parameter Values In Stochastic Kinetic Models

Murali Koneru, Michael Benton and **Martin A. Hjorts**, Chemical Engineering, Louisiana State University, Baton Rouge, LA

Stochastic biological models can describe a range of phenomena that are to a certain extend random, such as emergence of cancer cells, apoptosis or emergence of drug resistance. A concern with these models is the large number of model parameters that may not be easily determined by traditional least square techniques used for deterministic models.

In the present work, several stochastic models of the budding yeast cell cycle are presented and approaches to estimate the model parameters are discussed. We have found that in-silico evolution is a potentially useful tool for obtaining model parameter values that give biologically reasonable results. Using this algorithm, growth of a cell population is simulated such that the stochastic parameters are allowed to change in each division while the population size is kept constant by continuous removal of cells based on some selection criteria. The selection criteria can be as simple as removal of cells at random, in which case the algorithm selects for the faster growing phenotype.

The simulation results show that the average cell mass converges rapidly to a steady value while many of the stochastic parameters continue to fluctuate. Presumably, large fluctuations indicate that the value of a parameter is not important in determining the phenotype while small fluctuations indicate the opposite. The magnitude of the allowable change in parameter values at division affects the execution time of the algorithm and convergence to a steady value and allowing the changes to be too great results in excessive cell death.

See more of this Session: Modeling Approaches In the Life Sciences I
See more of this Group/Topical: Topical A: Systems Biology

(313g) Spatial Mechanisms for the Stepwise Navigation of Neutrophils

Yuki Kimura[1], Ashish Kapoor[2], Paul J. A. Kenis[1], Luke Olson[3] and Christopher V. Rao[2], (1)Chemical and Biomolecular Engineering, University of Illinois at Urbana-Champaign, Urbana, IL, (2)Chemical and Biomolecular Engineering, University of Illinois, Urbana Champaign, Urbana, IL, (3)Computer Science, University of Illinois at Urbana-Champaign, Urbana, IL

In response to infection, immune cells are recruited to fight off invading microbes. Chemotaxis, the process by which cells move in response to chemical gradients, plays a prominent role in this defense mechanism. A number of chemical signals, called chemoattractants, are produced at or proximal to sites of infection and diffuse into the surrounding tissue. Immune cells sense these chemoattractants and move in the direction where their concentration is greatest, thereby locating the source of attractants and the associated targets. Leading the assault against new infections is a specialized class of white blood cells called neutrophils, which normally circulate in the blood stream. Upon activation, neutrophils adhere to and squeeze through the vascular endothelium, crawling to sites of infection and inflammation. There they phagocytose bacteria and release a number of proteases and reactive oxygen intermediates with antimicrobial activity.

Neutrophils respond to many different chemoattractants. One open question is how they locate their targets in complex environments consisting of many different chemoattractant species and sources. Successful navigation under such conditions necessitates a robust mechanism for interpreting and prioritizing the signals received. Recent studies have implicated an intrinsic signaling hierarchy among the known chemoattractants, resulting in their classification as either end target or endogenous species. In particular, the end target variants, produced exclusively at or near the site of infection, have been found to consistently take precedence over the endogenous species. Endogenous chemoattractants, on the other hand, induce an apparently counter-intuitive chemotactic response within the cells, where distant sources are favored over proximal sources regardless of their type.

To date, several attempts to explain this preferential bias toward distant sources have involved temporal mechanisms such as sensory adaptation or an inherent delay in the cell response. However, the possibility of spatial effects has remained unaddressed. Here, we present a mathematical model of neutrophil chemotaxis to show that the observed bias can also be recovered by a purely spatial mechanism - the movement of the cell through changing environments naturally leads to changes in sensitivity to different concentrations of the chemoattractants. Moreover, we show that this mechanism enhances the ability of neutrophils to locate target infections within complex environments. This corroborates a recent hypothesis that neutrophils may migrate in a stepwise fashion between chemokine sources as a means to approach distant end targets. Finally, we present experimental results to validate the predictions of our model.

(313h) Dynamic Metabolic Flux Analysis (DMFA): A Framework for Determining Fluxes At Metabolic Non-Steady State

Maciek R. Antoniewicz, Chemical Engineering, University of Delaware, Newark, DE and Robert Leighty, Department of Chemical Engineering, Univerisity of Delaware, Newark, DE

Metabolic flux analysis (MFA) is a key tool for measuring *in vivo* metabolic fluxes in systems at metabolic steady state. Here, we present an extension to MFA for determining flux dynamics in systems at metabolic non-steady state. The dynamic MFA (DMFA) framework is based on non-stationary stoichiometric mass balances to describe flux transients. We derived exact algebraic solutions to the DMFA problem and developed efficient algorithms for least-squares parameter estimation and statistical analysis in these dynamic models. In DMFA, time-series of concentration measurements and external rate measurements are fitted in a single analysis to determine metabolic transients for the entire culture. Rigorous statistical criteria are employed to automatically identify characteristic metabolic phases. Here, we demonstrate the application of DMFA in three example systems. First, we evaluated the performance of DMFA in a simple three-reaction model in terms of accuracy, precision and flux observability. Next, we analyzed a commercial glucose-limited fed-batch process for 1,3-propanediol production. The DMFA methodology accurately captured the dynamic behavior of the fed-batch fermentation and identified characteristic metabolic phases. Lastly, we demonstrate the application of DMFA without an assumed metabolic model for detecting gross measurement errors in fermentation data and for data reconciliation using carbon and electron balances as constraints. The DMFA approach that we developed here is ideally suited for monitoring metabolic shifts in industrial fermentations.

See more of this Session: Modeling Approaches In the Life Sciences I
See more of this Group/Topical: Topical A: Systems Biology

(313i) Genome Scale Metabolic Reconstruction and Hypothesis Testing In the Methanogenic Archaeon Methanosarcina Acetivorans C2A

Matthew Nicholas Benedict[1], Matthew C Gonnerman[1], William W Metcalf[2] and Nathan D. Price[1], (1) Chemical and Biomolecular Engineering, University of Illinois at Urbana-Champaign, Urbana, IL, (2) Microbiology, University of Illinois at Urbana-Champaign, Urbana, IL

Methanosarcina acetivorans strain C2A is a marine methanogenic archaeon notable for its substrate utilization, genetic tractability, and novel energy conservation mechanisms. To help probe the implications of this organism's unique metabolism, we have constructed and manually curated a genome-scale metabolic model, iMB744, accounting for 744 of the 4540 predicted protein coding genes (16%) in the *M. acetivorans* genome. The reconstruction effort has identified key knowledge gaps and differences in peripheral and central metabolism between methanogenic species. Using flux balance analysis, the model quantitatively predicts wild type phenotypes and is 96% accurate in knockout lethality predictions compared to currently available experimental data. Flux balance analysis was used to probe the mechanisms and energetics of byproduct formation and growth on carbon monoxide, and the nature of the reaction catalyzed by the soluble heterodisulfide reductase HdrABC in *M. acetivorans.* This work highlights the great utility of constraint-based modeling for identifying feasible solutions to biological questions and provides insights into the workings of the cell at the genome scale.

See more of this Session: Modeling Approaches In the Life Sciences I
See more of this Group/Topical: Topical A: Systems Biology

(390a) Proteomic, Physiological, and *In Silico* Testing of Economic Tradeoffs In Metabolic Networks

Ross P. Carlson, **Reed Taffs** and James P. Folsom, Chemical and Biological Engineering, Montana State University, Bozeman, MT

Microbes conserve energy and perform carbon conversions using robust and often redundant networks of enzymatic machinery. Competition for resources in nature has likely selected for phenotypes wherein operational metabolic pathways optimize tradeoffs between catabolic efficiency of an enzyme set and the investment of anabolic nutrients required to synthesize the enzymes. An *in silico* economics-based analysis quantified molecular-level, resource allocation tradeoff strategies that permit competitive cellular functioning under a continuum of nutrient scarcities. The analysis decomposed a metabolic network into a complete listing of non-divisible, mathematically-defined biochemical pathways which were then used to identify all strategies for investing limiting resources like iron and nitrogen into the genome encoded metabolic machinery. Nitrogen, for example, is required for the synthesis of every enzyme; as nitrogen availability decreases, tradeoff analysis predicts a testable shift from catabolism optimized phenotypes to nitrogen investment-optimized phenotypes. *In silico* predictions were evaluated experimentally using physiological and proteomic data collected from iron- or nitrogen-limited *Escherichia coli* chemostat cultures. Experimental chemostat data trends were consistent with *in silico* theory and illustrated that under iron- and nitrogen-limited conditions *E. coli* regulates its metabolism to invest scarce resource competitively at the cost of optimal biomass yields on electron donor. The study highlights a fundamental evolutionary and metabolic design paradigm for competitive network structure and control with applications to bioprocess engineering, environmental microbiology and treatment of microbial pathogens.

See more of this Session: Modeling Approaches In the Life Sciences II
See more of this Group/Topical: Topical A: Systems Biology

(390b) Pattern Formation In the Acto-Myosin Cortex: An Active Reaction-Diffusion-Advection Process

Justin S. Bois[1], Frank Jlicher[2] and Stephan W. Grill[2], (1)Chemistry and Biochemistry, UCLA, Los Angeles, CA, (2)Biological Physics, Max-Planck Institute for the Physics of Complex Systems, Dresden, Germany

The acto-myosin cortex, which resides immediately below the eukaryotic cell membrane, is a thin network of crosslinked actin filaments on which myosin motors exert stress though active processes involving ATP hydrolysis. The cortex is continuously remodeled, so that on long time scales it may be treated as a viscous fluid. Regulators of myosin activity and actin dynamics, and therefore active stress, are also present in the cortex. To capture this biochemical and mechanical coupling, we have developed a general hydrodynamic description of the cortex as a thin film of an active viscous fluid in which active stresses are under regulation of biochemical species which undergo diffusion and chemical reactions. The coupling of active mechanical stresses with a reaction-diffusion system results in biochemical pattern formation with steady viscous flow. In many cases, active fluid motion promotes patterns in systems which would otherwise remain homogeneous. Such patterns are prevalent in embryonic development, cell motility, and polarization, and this theory provides a simple, general framework in which they may be understood. We apply the theory to the dynamics of the polarization of the one-cell *C. elegans* embryo, a system for which mechanical and chemical coupling is essential.

See more of this Session: Modeling Approaches In the Life Sciences II
See more of this Group/Topical: Topical A: Systems Biology

(390c) SHREDing a Biochemical Network Into Hierarchical Modules

Gautham V. Sridharan[1], Soha Hassoun[2] and **Kyongbum Lee**[1], (1)Chemical and Biological Engineering, Tufts University, Medford, MA, (2)Computer Science, Tufts University, Medford, MA

Hierarchical modularity has emerged as anorganizational principle of biochemical networks, providing insights into thecoordinated regulation of reactions within and across pathways. In principle, themodularity of a biochemical network should allow the system to be partitioned intominimally interdependent parts, which in turn can facilitate detailed analysisof each part in the context of the overall system. In practice, modularityanalysis has often relied on *ad hoc*decisions to mainly corroborate existing knowledge. While there is generalagreement that a module should consist of a biologically meaningful group ofconnected components in the network, there is little consensus on the metricneeded to quantitatively evaluate the quality of a partition. The goal of thisstudy is to investigate a novel metric that can be used to *systematically* partition a biochemical network into functionallyrelevant groups of reactions. The metric, termed the Shortest RetroactiveDistance (ShReD),characterizes the retroactive connectivity between any two reactions in anetwork arising from potential feedback interactions, thereby grouping togthernetwork components which mutually influence each other. We evaluate the metricusing two test networks: epidermal growth factor receptor (EGFR) signaling and liverdrug transformation.

Each test network was abstracted as areaction-centric directed. An edge was drawn betweentwo reaction nodes existed if one reaction produced either a reactant orallosteric effector of the other reaction. The ShReDbetween two reactions was calculated as the path-length of the shortest cycle thatinvolves both reaction nodes. Using ShReD as the measure of connectivity, weadapted Newman's algorithm[1] to obtain a set of hierarchical partitions.Briefly, Newman's original algorithm dividesan *undirected* network such that theresulting partition maximizes the number of connections *within* each sub-network *relativeto* the number of expected connections between two randomly chosen nodes inthe network. In this study, the partitions were performed based on the ShReD metric, rather than the undirected connectivity usedin Newman's work.

For the EGFR network, thepartitions obtained using Newman's original algorithm and ShReDwere largely similar (not shown). Importantly, the ShReDpartition generated reaction modules with larger numbers of cyclicalinteractions. For the metabolic network, which included cofactors, the ShReD partition again generated hierarchical modules whosecompositions compared favorably with canonical associations based on textbookbiochemistry (Fig. 1). The Newman partition was unable to generate anyhierarchy. Interestingly, the ShReD partitionrevealed a 'redox' module involving reactions of xenobiotic transformation,glucose metabolism, pyruvate metabolism, and lipid metabolism interactingthrough shared production and consumption of NADPH.

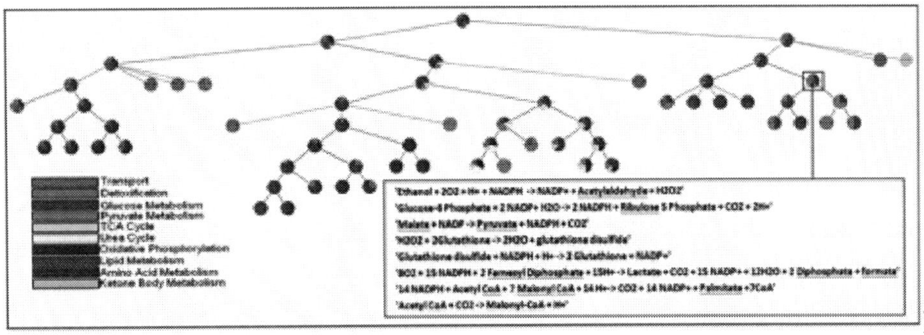

Fig. 1. ShReD-based hierarchical partition of a hepatocyte metabolic network.The pie charts represent modules of reactions. Each pie represents the fractionof

Inconclusion, our novel metric ShReD, combined with Newman's algorithm, is to ourknowledge the first modularity analysis technique that partitions a biochemicalnetwork to preserve cylical interactions between reactions.

1. Newman, ME (2006). Proc Natl Acad SciU S A. 103(23): 8577-82.

See more of this Session: Modeling Approaches In the Life Sciences II
See more of this Group/Topical: Topical A: Systems Biology

(390d) Shewanella Oneidensis MR-1 Population-Level Chemotaxis In a Diffusion Gradient Chamber

Rui Li[1], Jennifer M. Auchtung[2], James M. Tiedje[2] and R. Mark Worden[1], (1)Chemical Engineering and Materials Science, Michigan State Univeristy, East Lansing, MI, (2)Center for Microbial Ecology, Michigan State University, East Lansing, MI

Shewanella species are famous for their broad range of terminal electron acceptors, nearly ubiquitous presence in various environmental niches, and the ability to perform chemotaxis towards both organic and inorganic electron acceptors. Knowledge of *Shewanella's* chemotactic properties is important to understand its competitiveness and roles in elemental cycling and bioremediation (e.g., precipitation of soluble uranium oxides). Population-level microbial chemotaxis involves a complex interplay of cellular process (growth, metabolism, chemotaxis, and random motility) and molecular processes (diffusion of electron donors and acceptors that serve as chemoattractants). This paper describes use of a diffusion gradient chamber (DGC) to study population-level growth and chemotaxis of *Shewanella oneidensis* MR-1 in response to applied and cell-generated gradients of chemotactic electron acceptors. A mathematical model describing the key cellular and molecular processes governing *S. oneidensis'* chemotaxis was developed and used to analyze the experimental results. Population-level chemotactic responses of wild-type *S. oneidensis* MR-1 and electron-acceptor reductase mutants to uniformly distributed electron acceptors, pre-established unsteady or quasi-steady state gradient of electron acceptors were explored in the DGC. The model was able to reproduce key trends in the observed cell growth and migration patterns. The predicted velocities of chemotactic cell bands closely matched those measured in the DGC. These results validate the use of the DGC and its mathematical model to measure and simulate *Shewanella's* chemotaxis in response to electron acceptor gradients.

See more of this Session: Modeling Approaches In the Life Sciences II
See more of this Group/Topical: Topical A: Systems Biology

HIGHER ORDER GENE DELETION MUTANTS GIVE INSIGHTS ON LATENT PATHWAY ACTIVATION

Pratish Gawand, University of Toronto, Toronto, ON

Radhakrishnan Mahadevan, University of Toronto, Toronto, ON

It has been well established that deletion of metabolic genes causes a local redistribution of fluxes which results in a sub-optimal growth rate of the mutants (Segre *et al.*, 2002). However, the mutants can adaptively evolve to achieve the optimal FBA predicted growth rates (Fong and Palsson, 2004). Observations on temporal flux profiles of single gene deletion mutants have shown that latent pathways are transiently activated during the course of adaptive evolution (Fong *et al.*, 2006). The reason for the activation of these pathways, however, still remains speculative. Recently, it has been proposed that the presence of these pathways adversely affects the growth rates during the sub-optimal growth phase (Cornelius *et al.*, 2011). Furthermore, it is not well understood whether the activation of latent pathways is a suboptimal response to the genetic perturbations or whether it is a global non-specific response.

We attempted to understand the phenomenon using higher order gene deletion mutants, which showed different growth phenotypes on different carbon sources. Two *Escherichia coli* mutants, one with three gene deletions (LMSE2) and one with five gene deletions (LMSE5), were used in the study. The strategic gene deletions in the two mutants were expected to give insights on the latent pathways activation under different media conditions. Using the genome-scale model iAF1260, the latent pathways for both the mutants were identified as the reactions that were activated during the sub-optimal growth phase (predicted using MOMA) and but not active during optimal growth phase (predicted using FBA) (Cornelius *et al.*, 2011). Abilities of the mutants to recover on different carbon sources after gene deletions, was used to predict the activation of the latent pathways in these mutants.

Our experimental observations suggest that latent pathways are either not activated or are unable to support growth in an event of major rerouting of fluxes. In addition to this, our results indicate that the ability of the mutants to return to the original growth state plays a major role in the recruitment of latent pathways and hence, determines the mutant's survival. It has been previously suggested that mutants do not recover in the event of major flux re-routing. Our results are in agreement with this observation and show that in *E. coli*, the recovery of the mutant strongly correlates to the ability of the mutants to restore the original flux distribution.

References

1. Segrè, D., Vitkup, D. & Church, G. M. (2002) "Analysis of optimality in natural and perturbed metabolic networks", *Proc. Natl. Acad. Sci. U. S. A.* **99**, 15112-15117.

2. Fong, S. S. & Palsson, B. Ø. (2004) "Metabolic gene-deletion strains of Escherichia coli evolve to computationally predicted growth phenotypes", *Nat. Genet.* **36**, 1056-1058.

3. Fong, S. S., Nanchen, A., Palsson, B. O. & Sauer, U. (2006) "Latent pathway activation and increased pathway capacity enable Escherichia coli adaptation to loss of key metabolic enzymes", *J. Biol. Chem.* **281**, 8024-8033.

4. Cornelius, S., Lee, J. & Motter, A. (2011) "Dispensability of Escherichia coli's latent pathways", *Proc. Natl. Acad. Sci. U. S. A.* **108**, 3124-3129.

(390f) Derivation of a Kinematic Model Coupling Stress Fiber Dynamics with JNK Activation In Response to Matrix Stretching

Roland Kaunas, Department of Biomedical Engineering, Texas A&M University, College Station, TX, **Zuyi (Jacky) Huang**, Department of Chemical Engineering, Villanova University, Villanova, PA and Juergen Hahn, Department of Chemical Engineering, Texas A&M University, College Station, TX

Adherent cells in many tissues, including arteries, bladder, muscle, and tendon, to name a few, are subjected to stretch as the tissue deforms in response to external forces. Mechanical stretch regulates many cellular functions, including proliferation, apoptosis, migration and morphology that occur in response to stretch-induced changes in intracellular signaling and gene expression (Haga et al., 2007). While the mechanisms that regulate mechanotransduction remain unclear, focal adhesions and the actin cytoskeleton clearly play a role (Bershadsky et al., 2003). It is well established that focal adhesions and stress fibers are interdependent. This is supported by the fact that focal adhesions anchor stress fibers to the matrix and the integrity of the actin cytoskeleton is necessary to maintain focal adhesions (Volberg et al., 1994). Integrins have been implicated as mechanosensors responsible for stretch-induced intracellular signaling (Thodeti et al., 2009). Katsumi et al. (2005) reported that c-jun N-terminal kinase (JNK) activation in response to stretch requires the formation of new integrin–matrix bonds. We have shown that cells, along with their stress fibers and focal adhesions, orient perpendicular to the direction of cyclic uniaxial stretch (Kaunas et al., 2005). The time course of cyclic stretch-induced JNK activation corresponds with that of stress fiber realignment (Kaunas et al., 2006). Specifically, cyclic uniaxial stretch causes the transient activation of JNK that subsides as the stress fibers become oriented perpendicular to the direction of stretch. Stretch-induced stress fiber alignment occurs via disassembly and reassembly of stress fibers (Hayakawa et al., 2001), which would result in the formation of new integrin–matrix bonds at the associated focal adhesions. Together, these studies suggest that stretch-induced JNK activation is upregulated during times when stress fibers and their associated integrin–matrix adhesions are undergoing relatively high rates of disassembly and reassembly. It is of interest to investigate intracellular signaling that is associated with JNK activation in response to different patterns of stretch stimulation using a mathematical modeling approach, as it can provide systems-level insight into the complex relationships between matrix stretch pattern, stress fiber dynamics, integrin turnover, and JNK signaling.

In order to link the kinetics of intracellular signaling with different patterns of stretch, a model is needed that describes the temporal and spatial evolution of stress fiber organization in response to stretch. Recently, computational modeling has been used to elucidate the complex interplay between matrix deformation, cytoskeletal and integrin bond dynamics and stress fiber reorganization. For example, Kaunas and Hsu (2009) have formulated a computational model of stress fiber dynamic reorganization based on constrained mixture theory (Humphrey, 2008) that successfully describes published time courses of stress fiber reorientation in response to cyclic uniaxial and equibiaxial stretches. Based on the interdependence of stress fibers, integrins, and JNK activation, we coupled the stress fiber dynamics model presented by Kaunas and Hsu (2009) to a kinetic model of JNK activation in this work. Specifically, we developed a mathematical model coupling the dynamic disassembly and reassembly of actin stress fibers and associated focal adhesions to the activation of JNK in cells attached to deformable matrices. The model is based on the assumptions that stress fibers are pre-extended to a preferred level under static conditions and that perturbations from this preferred level destabilize the stress fibers. Numerical solutions of the developed model equations predict that different patterns of matrix stretch result in distinct temporal patterns in JNK activation that compare well with published experimental results. In the case of cyclic uniaxial stretching, stretch-induced JNK activation slowly subsides as stress fibers gradually reorient perpendicular to the stretch direction. In contrast, JNK activation is chronically elevated in response to cyclic equibiaxial stretch. A step change in either uniaxial or equibiaxial stretch results in a short, transient upregulation in JNK that quickly returns to the basal level as overly stretched stress fibers disassemble and are replaced by fibers assembled at the preferred level of stretch. In summary, the model describes a mechanism by which the dynamic properties of the actin cytoskeleton allow cells to adapt to applied forces through turnover and reorganization to modulate intracellular signaling.

References

Bershadsky, A.D., Balaban, N.Q., Geiger, B., 2003. Adhesion-dependent cell mechanosensitivity. Annu. Rev. Cell Dev. Biol. 19, 677–695.

Haga, J.H., Li, Y.S., Chien, S., 2007. Molecular basis of the effects of mechanical stretch on vascular smooth muscle cells. J. Biomech. 40, 947–960.

Hayakawa, K., Sato, N., Obinata, T., 2001. Dynamic reorientation of cultured cells and stress fibers under mechanical stress from periodic stretching. Exp. Cell Res. 268, 104–114.

Humphrey, J.D., 2008. Vascular adaptation and mechanical homeostasis at tissue, cellular, and sub-cellular levels. Cell Biochem. Biophys. 50, 53–78.

Katsumi, A., Naoe, T., Matsushita, T., Kaibuchi, K., Schwartz, M.A., 2005. Integrin activation and matrix binding mediate cellular responses to mechanical stretch. J. Biol. Chem. 280, 16546–16549.

Kaunas, R., Hsu, H.J., 2009. A kinematic model of stretch-induced stress fiber turnover and reorientation. J. Theor. Biol. 257, 320–330.

Kaunas, R., Usami, S., Chien, S., 2006. Regulation of stretch-induced JNK activation by stress fiber orientation. Cell Signal. 18, 1924–1931.

Kaunas, R., Nguyen, P., Usami, S., Chien, S., 2005. Cooperative effects of Rho and mechanical stretch on stress fiber organization. Proc. Nat. Acad. Sci. USA 102, 15895–15900.

Thodeti, C.K., Matthews, B., Ravi, A., Mammoto, A., Ghosh, K., Bracha, A.L., Ingber, D.E., 2009. TRPV4 channels mediate cyclic strain-induced endothelial cell reorientation through integrin-to-integrin signaling. Circ. Res. 104, 1123–1130.

Volberg, T., Geiger, B., Citi, S., Bershadsky, A.D., 1994. Effect of protein kinase inhibitor H-7 on the contractility, integrity, and membrane anchorage of the microfilament system. Cell Motil. Cytoskeleton 29, 321–338.

See more of this Session: Modeling Approaches In the Life Sciences II
See more of this Group/Topical: Topical A: Systems Biology

(390g) A Systems Biology Approach to Identify Immune Targets That Control Tuberculosis Granuloma Function

Mohammad Fallahi-Sichani[1], Denise E. Kirschner[2] and Jennifer J. Linderman[1], (1)Chemical Engineering, University of Michigan, Ann Arbor, MI, (2)Microbiology and Immunology, University of Michigan Medical School, Ann Arbor, MI

Tuberculosis (TB) is a common and deadly infectious disease caused by a highly successful bacterium, *Mycobacterium tuberculosis* (Mtb). Multiple host immune factors control the formation of a self-organizing aggregate of immune cells termed a granuloma, after inhalation of Mtb. One such factor, tumor necrosis factor-α (TNF), is a protein that regulates inflammatory immune responses. TNF (in conjunction with the cytokine IFN-γ) induces macrophage activation, enhances immune cell recruitment to the site of infection, and augments chemokine and cytokine expression by macrophages through activation of the NF-κB signaling pathway. However, there are differences among dynamics of each of these NF-κB-associated responses as a result of differences in molecular processes (e.g. RNA and protein degradation rates) acting downstream of TNF-induced NF-κB activation. The impact of these dynamical differences on formation and function of granulomas and thus the outcome of Mtb infection is not known.

We have developed a multi-scale computational model that describes the immune response to Mtb in lung over three biological length scales: tissue, cellular and molecular. Using this model, we predict the impact of NF-κB associated response dynamics on the outcome of infection at the level of a granuloma. Our model suggests that the timing of these responses, in addition to the extent of response, plays a critical role in control of infection and inflammation. Manipulations of the dynamics of these responses lead to different outcomes, including clearance of bacteria, containment of bacteria within a stable granuloma, uncontrolled growth of bacteria, or excessive inflammation. The intracellular NF-κB associated signaling molecules and processes involved in TB immunity that we identify as crucial may be new targets for therapy.

See more of this Session: Modeling Approaches In the Life Sciences II
See more of this Group/Topical: Topical A: Systems Biology

(390h) Dissecting Gene Regulation Through DNA Loop Formation Using Statistical Mechanical Models and Experiments

James Boedicker, Applied Physics, Caltech, Pasadena, CA, Hernan Garcia, Physics, Caltech, Pasadena, CA and Rob Phillips, Applied Physics and Mechanical Engineering, Caltech, Pasadena, CA

Testing ourunderstanding of genetic circuits requires models which make quantitativepredictions that can be experimentally validated. A common motif of generegulation involves the interactions between transcription factors, RNApolymerase, and DNA. Although many examples of gene regulation through suchinteractions are well known, our understanding of even the simplest suchsystems is often qualitative, at the level of biological °Cartoons. Thesecartoons only predict whether interactions between the regulatory components increase,decrease, or have no effect on gene expression. Such models fail to makequantitative predictions about specific realizations of such genetic circuits, evenfor well-studied systems such as the lac operon.

Our approach hasbeen to construct models of gene expression based on equilibrium statisticalmechanics. The models require only a few input parameters. The modelparameters include binding energies of proteins, copy numbers of components,and the mechanical properties of DNA. These parameters are measured in simple controlexperiments. Once these inputs have been determined, the models contain nofree parameters and make quantitative predictions about the level of geneexpression over a wide range of parameter space. We have constructed andexperimentally validated such models to test our understanding of the lacoperon.

In the lac operon,a dimeric transcription factor, lac repressor, can bind to three differentbinding sites on DNA, called operators. Gene expression is reduced byrepressor binding to the center operator just adjacent to the promoter. It isthought that binding to the outer, auxiliary sites alone cannot directlyinfluence gene expression. Instead, the auxiliary sites can participate inloop formation, in which one dimeric repressor can simultaneously bind to boththe center and an auxiliary operator, bending the intervening DNA into a loop.The looped state leads to enhancement of repression due to the loop locking inthe repressor on the center operator.

We use thisgenetic circuit to make predictions about how the specifics features of the lacoperon lead to a specific level of gene expression. The model we developed forthe lac operon has six states, with the probability of each state weighted by aBoltzmann factor calculated from the change in free energy associated with eachstate. Some of the states lead to repression while others enable binding ofthe RNA polymerase and transcription of the gene. Using the model, we makequantitative predictions about how the level of gene expression is modulated byeach component of the system. We model and test how changes in the bindingenergies of the repressor to the operators, the number of repressors per cell, thelength of the DNA loop, and the flexibility of the DNA in the loop alter thelevel of gene expression.

The models weretested experimentally by creating genetic circuits containing the generegulatory components of the lac operon. These circuits were incorporated intothe genome of *E. coli*. Fluorescent reporters were used to quantify thelevel of gene expression in each circuit. The readout of the experiments wasthe absolute level of fluorescent protein. Using this approach, thepredictions of the thermodynamic equilibrium models were experimentally testedfor many genetic circuits containing different combinations of the regulatoryelements of the lac operon.

The experiments revealthat the standard model described above does not hold. There is a contributionfrom the downstream auxiliary operator to repression, even in the absence ofloop formation. This result shows that binding events downstream of thepromoter can still interfere with the activity of RNA polymerase, and do so ina manner which is dependent upon the orientation of the promoter and operator.Additionally, the intrinsic flexibility of the looping DNA has no effect on thelevel of repression. DNA sequences that have previously been shown to easilyform loops *in vitro* do not lead to increased repression. The apparentlack of dependence of repression on the flexibility of the looping DNA suggeststhat other *in vivo* factors such as DNA bending proteins are dominant indetermining the probability of the looping in vivo. These results highlightthe ability of quantitative models of gene expression, such as the zero free parameterthermodynamic equilibrium model applied here, to test and expand our understandingof gene regulatory mechanisms. This approach should lead to both mechanisticinsights into gene regulation *in vivo*, and also enable the constructionof synthetic genetic circuits with fine-tuned control of gene expression.

See more of this Session: Modeling Approaches In the Life Sciences II
See more of this Group/Topical: Topical A: Systems Biology

(390i) Regulation of Epidermal Growth Factor Receptor Phosphorylation and Endocytosis Dynamics by Protein Tyrosine Phosphatases

Calixte S. Monast and Matthew J. Lazzara, Chemical and Biomolecular Engineering, University of Pennsylvania, Philadelphia, PA

Cell signaling pathways initiated by the epidermal growth factor receptor (EGFR) are broadly involved in normal physiology and pathology. Phosphorylated EGFR levels, which dictate receptor signaling strength and persistence, are the net result of two opposing processes, auto-phosphorylation of cytoplasmic tyrosines by the receptor kinase and dephosphorylation of those tyrosines by protein tyrosine phosphatases (PTPs). Whereas the processes leading to tyrosine phosphorylation have been studied in great detail, relatively little is known about the kinetics of EGFR tyrosine dephosphorylation or the coupling of this process to other receptor-level phenomena which depend upon receptor phosphorylation, such as endocytosis. Here, we developed a novel mechanistic model of EGFR phosphorylation dynamics to determine the rates of EGFR dephosphorylation at the plasma membrane and in the cell interior and applied the model to generate predictions for the effects of these dephosphorylation kinetics on EGFR endocytosis dynamics. Our model was constructed to allow for the possibility of different rates of receptor dephosphorylation at the plasma membrane and cell interior and accounted for interactions of EGFR with itself (dimerization and phosphorylation), ATP, kinase inhibitors, PTPs, and adapter proteins involved in receptor trafficking processes. To determine rate constants for EGFR tyrosine dephosphorylation at the plasma membrane and cell interior, the model was regressed against a diverse set of experimental data for receptor phosphorylation response to EGFR ligands, PTP inhibitors, and EGFR kinase inhibitors as well as data for steady EGFR phosphorylation levels in the absence of ligands. Our regression results led to the novel conclusion that EGFR dephosphorylation is significantly more rapid at the plasma membrane than in the cell interior, a finding which contradicts the classical view of EGFR regulation by PTPs and several previous computational studies of EGFR phosphorylation dynamics which were developed by considering data for receptor response to ligand alone. In fact, our model results suggest that the process of EGFR dephosphorylation is so rapid that receptor tyrosines may go through a complete phosphate cycle (phosphorylation and dephosphorylation) as many as thirteen times per minute. Thus, our model results suggest that, as with other kinase-phosphatase systems, EGFR and its associated PTPs participate in a "futile cycling" process wherein phosphate cycling occurs on a time scale much smaller than that for other important receptor processes such as endocytosis and degradation. Since the most dominant pathways for EGFR endocytosis require phosphorylation of key EGFR tyrosines, a feature encoded within our model, our results further suggest that regulation of EGFR by PTPs at the plasma membrane may play a significant role in determining rates of ligand-mediated EGFR endocytosis. In fact, for an EGFR expression level characteristic of epithelial cells, our model predicted that PTP activity at the plasma membrane may reduce the rate constant for ligand-mediated EGFR endocytosis by up to six-fold from the value it would otherwise be in the absence of PTP activity at the plasma membrane. This predicted effect of PTPs on EGFR endocytosis could be reduced by the inclusion of the trafficking adapter protein referred to previously in the model calculations. Experimental measurements of the effects of PTPs on EGFR endocytosis were made for cultured fibroblasts and epithelial cells using radiolabeled EGFR ligands or antibodies and inhibitors of PTP activity. Overall, comparison between our measurements and model calculations suggested that PTP activity at the plasma membrane influences EGFR endocytosis kinetics but that the enhanced binding of adapter proteins in the context of PTP inhibition reduces the observed effect from the maximum predicted effect.

See more of this Session: Modeling Approaches In the Life Sciences II
See more of this Group/Topical: Topical A: Systems Biology

(534a) Identifying Perturbed Pathways In Glioblastoma Through Network Reconstruction and Analysis

James A. Eddy, Bioengineering, University of Illinois at Urbana Champaign, Urbana, IL and Nathan D. Price, Chemical and Biomolecular Engineering, University of Illinois at Urbana-Champaign, Urbana, IL

Introduction: Perturbations to biomolecular networks in glioblastoma multiforme (GBM)—the most common and devastating form of brain cancer—eventually result in the symptoms of disease observed by the patient, including seizure, nausea, vomiting, headache, and a progressive deterioration of memory and personality. These phenotypes arise not from any single mutation, but from the sum effect of complex interactions among multiple aberrant genes and other molecular agents. The combinatorial nature of GBM tumor development therefore mandates a systems-level approach to elucidate underlying mechanisms of the cancer. Reconstructing detailed *in silico* models of biochemical reaction networks (e. g., metabolic, signaling, regulatory) at the genome scale establishes a platform on which genetic perturbations can be related to emergent malignant functions and phenotypes. We have reconstructed the first genome-scale network of glioblastoma metabolism, as altered function in several metabolic pathways has been shown to be critical in the development of GBM as well as other cancers [1]. Constraint-based modeling with the network allows us to simulate cell growth under physiologically relevant conditions and investigate mechanisms for tumor development. Curated pathways in the reconstructed network also provide functional context for statistical approaches to study gene expression in disease states. These combined approaches enable the identification and detailed characterization of key perturbed pathways in GBM.

Materials and Methods: Utilizing the recent generic human metabolic reconstruction—a well-curated repository of reaction information—we have computationally integrated gene and protein expression data from multiple sources to define a network specific to GBM. Specifically, we first used the Model Building Algorithm (MBA) [2] to systematically prune reactions from *Human Recon 1* [3]. Physico-chemical and environmental constraints were then applied to interrogate properties of the reconstructed network *in silico*, and to simulate cell growth under physiologically relevant conditions. Experimental measurements from the U87 GBM cell line—including metabolomics, siRNA knockdown phenotypes, and growth rates under various media conditions—were used, in conjunction with simulations, to refine, expand, and validate the model. To study pathway-level changes in gene expression relevant to GBM, we employed a method previously developed in our lab. Differential Rank Conservation (DIRAC) [4] was used to assess how pathway rankings (i.e., the relative expression ordering among genes within model pathways) differ, either among GBM samples, or between tumors and healthy astrocytes.

Results and Discussion: The model represents the subset of human metabolic capability predicted to be active in GBM, based on tissue-specific gene and protein expression, as well as experimental and literature data. Using the model, we were able to accurately predict experimental values we observed for U87 cell lines, including growth rates and lethality of gene deletions. Furthermore, the model is able to capture known metabolic phenotypes in cancer cells, including the Warburg effect. We examined GBM metabolic pathways with DIRAC and identified networks that are (i) tightly regulated within tumors and astrocytes, as defined by high conservation of transcript ordering; or (ii) variably expressed (i.e., shuffled) between phenotypes, resulting in specific classifiers that predict cancer versus healthy with high accuracy.

Conclusions: Cancer is an intrinsically complex and heterogeneous disease, making it particularly amenable to systems biology approaches. Using the mechanistically-detailed model we have built for GBM, we can begin to interrogate the link between mutations in the cancer and key metabolic processes that contribute to tumorigenesis. The model also serves as a powerful tool for studying pathway-level gene expression changes in GBM; perturbed pathways identified by DIRAC represent robust differences between healthy and diseased states, and in turn, serve as focal points for ongoing model simulation and development. In the future, systems-level analysis of GBM with the model should continue to provide important biological insights into underlying mechanisms and potential treatment routes for the disease.

References:

1. Cairns, RA, et al., Nat Rev Cancer, 2011, 11:85-95.
2. Jerby, L, et al., Mol Syst Biol, 2010, 6:401.
3. Duarte, NC, et al., P Nalt Acad Sci USA, 2007, 104:1777-82
4. Eddy, JA, et al., PLoS Comput Biol, 2010, 6:e1000792.

See more of this Session: In Silico Systems Biology: Cellular and Organismal Models
See more of this Group/Topical: Topical A: Systems Biology

(534b) Peripheral Clock Gene Entrainment by Cortisol

Panteleimon D. Mavroudis[1], Jeremy D. Scheff[2], Steven E. Calvano[3], Stephen F. Lowry[3] and Ioannis P. Androulakis[4], (1)Chemical and Biochemical Engineering, Rutgers University, Piscataway, NJ, (2)Biomedical Engineering, Rutgers University, Piscataway, NJ, (3)Department of Surgery, UMDNJ-Robert Wood Johnson Medical School, New Brunswick, NJ, (4)Biomedical Engineering, Chemical & Biochemical Engineering, Rutgers University, Piscataway, NJ

Peripheral Clock Gene Entrainment by Cortisol

The objective of this study is to assess the entrainment of peripheral bloodleukocyte (PBL) clock genes by cortisol and evaluate its entrainingcharacteristics. We demonstrate that the circadian characteristics of theperipheral entrainer (cortisol) have significant implications on the dynamicsof clock genes. In the presence of a normal circadian pattern in cortisol, the PBLsare entrained to a coherent diurnal pattern. As the amplitude of cortisol'scircadian rhythm is decreased slightly, the PBL oscillators remain entrainedbut undergo a phase shift relative to cortisol's phase. And as the amplitude isfurther decreased, eventually the PBLs fall out of sync and produce a flatensemble average. Understanding the relationship between the pattern ofcortisol secretion and its downstream effects on peripheral clock genes lays afoundation for studying the implications of dysregulated cortisol secretion(most notably under stress) on the function of PBLs.

To study the effect of cortisol in peripheral clock genes, we propose aunified model involving the circadian production of cortisol, cortisol signaltransduction, and clock gene oscillations. We explore the "two rates" model ofcircadian cortisol production (Chakraborty, Krzyzanski et al. 1999) to generate the diurnal pattern in plasma cortisol concentration by switching between a high secretion rate during the morning when cortisol concentration is increasing and a low secretion rate the rest of the day when cortisol concentration decreases to a low resting level. Cortisol in plasma moves into the cytoplasm of PBLs, where it binds to glucocorticoid receptor (GR), forming an activated drug-receptor complex which subsequently translocates into the nucleus and acts as a transcription factor, regulating a wide range of genes (Yao, DuBois et al. 2006).

We further hypothesize that the peripheral circadian gene network involves alimited set of critical genes, including Per1-3 and Cry1-2, which, incollaboration with the CLOCK/BMAL1 heterocomplex, form an orchestrated feedbackloop that leads to a circadian rhythm. Since Per1 has a glucocorticoidresponsive element (GRE) in its promoter region and is known to be responsiveto glucocorticoids (Yamamoto, Nakahata et al. 2005), we hypothesize that the primary link between cortisol's circadian rhythms and PBL circadian rhythms is the transcriptional activation of Per1 by the cortisol-GR activated complex. This circadian model includes an equation representing Per/Cry gene transcripts, which was modified by including an additive term in the Per/Cry equation representing stimulus of Per1 transcription by activated GR.

The purpose of our modeling effort is to evaluate the implications ofalterations in the circadian patterns of cortisol release in the entrainment ofclock genes. Our results demonstrate that under homeostatic release ofcortisol, robust entrainment of clock genes is observed, as well as fullsynchronization of the ensemble of PBLs. As the entraining characteristics ofcortisol are dysregulated, through a reduction in the amplitude of thecircadian pattern (a hallmark of stress), individual cells' clock genes fallout of phase until eventually they become desynchronized. The loss ofsynchronization is hypothesized to play a significant role in the attenuatedcontrol effects clock genes exert on the homeostatic response.

Preliminarymodel predictions are depicted in Figure 1. Ourmodel demonstrates the ability of cortisol's circadian variability tosynchronize clock genes ($R_{syn}= 1$) in an amplitude-dependentmanner. It predicts that a population of cells falls out of sync in the absenceof a systemic cortisol cue, whereas a circadian cortisol rhythm inducessynchronization of PBL clock genes. Furthermore, as the entrainer (cortisol)loses its circadian rhythmicity, our model predicts a regime of phase shift inperipheral clock genes relative to cortisol before synchronization is lost atvery low circadian amplitudes. Similar phase shifted responses have beenobserved in murine adipose tissues, where restricted feeding provokes adisruption of glucocorticoid rhythmicity along with phase shifts in theexpression of peripheral clock genes (Zvonic, Ptitsyn et al. 2006).

Cortisol rhythmicity plays animportant role in entraining peripheralclock genes and affecting the dynamics of the peripheral clock network. Ourwork has significant clinical implications as cortisol is a key component of thestress response and regulates the transcription of anti-inflammatory genes. Inhuman endotoxemia experiments, cortisol is acutely elevated and peripheralclock genes in PBLs are suppressed (Haimovich, Calvano et al. 2010), which may have important effects on the inflammatoryresponse, given the interplay between clock genes and cytokine production inperipheral immune cells (Coogan and Wyse 2008). Furthermore, in chronic stress when circadian rhythms in cortisol canbe lost (Lowry and Calvano 2008), peripheral circadian rhythms may be similarly desynchronized orshifted as predicted by our model. The work presented here will serve as afoundation for studying these interconnections between circadian rhythms andthe inflammatory response.

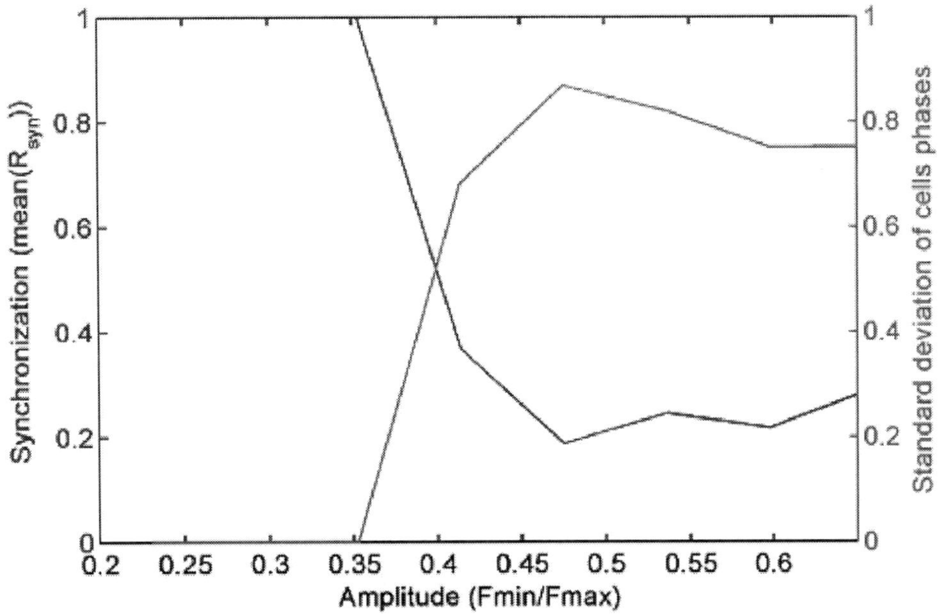

Figure 1: The degree of synchronization R_{syn}(ratio of the variance of the mean field to the mean variance of eachoscillator, $R_{syn}=1$ is synchronized, 0 otherwise) and the standarddeviation of the distribution of cells' phases both change as a function ofcortisol's amplitude.

References

Chakraborty, A., W. Krzyzanski, et al. (1999). "Mathematicalmodeling of circadian cortisol concentrations using indirect response models:comparison of several methods." J Pharmacokinet Biopharm **27**(1):23-43.

Coogan, A. N.and C. A. Wyse (2008). "Neuroimmunology of the circadian clock." BrainRes **1232**: 104-12.

Haimovich, B.,J. Calvano, et al. (2010). "In vivo endotoxin synchronizes and suppressesclock gene expression in human peripheral blood leukocytes." Crit CareMed **38**(3): 751-8.

Lowry, S. F.and S. E. Calvano (2008). "Challenges for modeling and interpreting thecomplex biology of severe injury and inflammation." J Leukoc Biol **83**(3):553-7.

Yamamoto, T.,Y. Nakahata, et al. (2005). "Acute physical stress elevates mouse period1mRNA expression in mouse peripheral tissues via a glucocorticoid-responsiveelement." J Biol Chem **280**(51): 42036-43.

Yao, Z., D. C.DuBois, et al. (2006). "Modeling circadian rhythms of glucocorticoidreceptor and glutamine synthetase expression in rat skeletal muscle." PharmRes **23**(4): 670-9.

Zvonic, S., A.A. Ptitsyn, et al. (2006). "Characterization of Peripheral CircadianClocks in Adipose Tissues." Diabetes **55**(4): 962-970.

See more of this Session: In Silico Systems Biology: Cellular and Organismal Models
See more of this Group/Topical: Topical A: Systems Biology

(534c) Towards Genome-Scale Dynamic Modeling of Cellular Metabolism. the Cybernetic Approach

Hyun-Seob Song, School of Chemical Engineering, Purdue University, West Lafayette, IN and Doraiswami Ramkrishna, Chemical Engineering, Purdue University, West Lafayette, IN

The hallmark of the cybernetic approach is its description of cellular regulation to provide for a dynamic framework of metabolic modeling. It views cells as optimal strategists frugally allocating internal resources among reactions in order to maximize a metabolic objective (e.g., carbon uptake rate or growth rate). Earlier development of cybernetic modeling had focused on the growth pattern on multiple substrates based on gross metabolic networks (e.g., Baloo and Ramkrishna 1991, Kompala et al. 1986, Ramakrishna et al. 1996, Turner et al. 1988). More recently, the hybrid cybernetic model (HCM) (Kim et al. 2008, Song et al. 2009, Song and Ramkrishna 2009) and lumped HCM (L-HCM) (Song and Ramkrishna 2010, 2011) has enabled consideration of more detailed networks through decomposition into elementary modes (EMs). Extension of this methodology to genome-scale networks, has, however, been challenged by difficulties associated with decomposition of large networks into EMs (Klamt and Stelling 2002). So far, genome-scale networks have been handled only by flux balance analysis (FBA) and its derivatives (Orth et al. 2010). Application of those constraint-based approaches to metabolic modeling and metabolic engineering is subject to limitations because of lack of attention to regulatory dynamics.

In this work, we present a methodology to incorporate genome-scale metabolic networks into the L-HCM framework. The L-HCMs describe cellular resources in terms of competition among lumped EMs (L-EMs) which are weighted averages of EMs in families classified according to similarity of metabolic function. As the weight takes a power-law form with an exponent of large constant, only a limited number of EMs play a role in computing L-EMs. Thus, instead of endeavoring to get the full set of EMs, we extract only these "essential" modes with an appreciable contribution to the averaging process. For this purpose, we have developed a MATLAB code based on the recursive mixed integer linear programming (MILP) algorithm originally proposed by Lee et al. (2000). In various test examples, it is shown that the MILP code successfully smokes out essential pathways including multiple optima and suboptima which are essential to compute L-EMs. This development enables the cybernetic modeling approach to address genome-scale networks. Also possible is direct comparison of L-HCMs with constraint-based approaches using the same size of genome-scale network.

References:

Baloo S, Ramkrishna D. 1991. Metabolic-Regulation in Bacterial Continuous Cultures .2. Biotechnology and Bioengineering 38(11):1353-1363.

Kim JI, Varner JD, Ramkrishna D. 2008. A Hybrid Model of Anaerobic E. coli GJT001: Combination of

Elementary Flux Modes and Cybernetic Variables. Biotechnology Progress 24(5):993-1006.

Klamt S, Stelling J. 2002. Combinatorial complexity of pathway analysis in metabolic networks. Molecular Biology Reports 29(1-2):233-236.

Kompala DS, Ramkrishna D, Jansen NB, Tsao GT. 1986. Investigation of Bacterial-Growth on Mixed Substrates - Experimental Evaluation of Cybernetic Models. Biotechnology and Bioengineering 28 (7):1044-1055.

Lee S, Phalakornkule C, Domach MM, Grossmann IE. 2000. Recursive MILP model for finding all the alternate optima in LP models for metabolic networks. Computers & Chemical Engineering 24(2-7):711-716.

Orth JD, Thiele I, Palsson BO. 2010. What is flux balance analysis? Nature Biotechnology 28(3):245-248.

Ramakrishna R, Ramkrishna D, Konopka AE. 1996. Cybernetic modeling of growth in mixed, substitutable substrate environments: Preferential and simultaneous utilization. Biotechnology and Bioengineering 52(1):141-151.

Song HS, Morgan JA, Ramkrishna D. 2009. Systematic Development of Hybrid Cybernetic Models: Application to Recombinant Yeast Co-Consuming Glucose and Xylose. Biotechnology and Bioengineering 103(5):984-1002.

Song HS, Ramkrishna D. 2009. Reduction of a Set of Elementary Modes Using Yield Analysis. Biotechnology and Bioengineering 102(2):554-568.

Song HS, Ramkrishna D. 2010. Prediction of Metabolic Function From Limited Data: Lumped Hybrid Cybernetic Modeling (L-HCM). Biotechnology and Bioengineering 106(2):271-284.

Song HS, Ramkrishna D. 2011. Cybernetic Models Based on Lumped Elementary Modes Accurately Predict Strain-Specific Metabolic Function. Biotechnology and Bioengineering 108(1):127-140.

Turner BG, Ramkrishna D, Jansen NB. 1988. Cybernetic Modeling of Bacterial Cultures at Low Growth-Rates - Mixed-Substrate Systems. Biotechnology and Bioengineering 32(1):46-54.

See more of this Session: In Silico Systems Biology: Cellular and Organismal Models
See more of this Group/Topical: Topical A: Systems Biology

(534d) A General Framework for Modeling Growth and Division of Mammalian Cells

John H. Gauthier, Complex Systems for National Security, Sandia National Laboratories, Albuquerque, NM and **Phillip I. Pohl**, Risk & Reliability Analysis,, Sandia National Laboratories, New Mexico, Albuquerque, NM

A framework for modeling complex cell-biologic processes is presented, based on two constructs: one describing the entire lifecycle of a molecule and the second describing the basic cellular machinery. Use of these constructs allows models to be built in a straightforward manner that fosters rigor and completeness. To demonstrate the framework, an example model of the mammalian cell cycle is presented that consists of several hundred differential equations of simple mass action kinetics. The additional complexity afforded by the framework allows the example model to be calibrated to both large-scale and small-scale observations and allows predictions to be made at both the systems level and the molecular level.

The example model allows predictions at both the systems level and the molecular level, including the following.

- When redundant pathways can activate a time-critical process, one preferred pathway must inactivate the other redundant pathway(s).
- Explicit growth monitoring is unnecessary during the mammalian cell cycle.
- Pre-translation mRNA regulation (other than splicing and the number of RNA polymerase) is unnecessary during the cell cycle.
- Transcription and translation (i.e., growth) continues during S phase at unimpeded rates.
- For cells capable of division, a ready supply of RNA polymerase must be available during G0.
- For cells capable of division, either a supply of DNA polymerase must be kept inactive during G0 or intense transcription and translation activity must occur and be controlled at a saturation level.
- SCF progressively complexes with Fbw7, Skp2, and Btrc through the cell cycle because of their differences in affinity with SCF and because of autoubiquitination in the absence of substrates.
- Cdc25A and Cdc25B cooperate to instigate the cycB/Cdk1-Cdc25C cascade and the G2/M transition.
- Transcription factors act multiplicatively.

This work also indicates a need for a method to determine the most robust set of rate parameters in a model of coupled ordinary differential equations.

See more of this Session: In Silico Systems Biology: Cellular and Organismal Models

(534e) Phenotype Prediction Through Genome-Scale Modeling and Raman Spectroscopy

Ryan S. Senger, Biological Systems Engineering Department, Virginia Tech, Blacksburg, VA

A cellular phenotype is determined by its metabolic network and the response of that network to environmental conditions. This means that both genetic and environmental perturbations can be responsible for altering the chemical composition of a cell. This has profound implications with microorganisms as we seek to understand the response of cultures to fermentive products (e.g., tolerance to biofuels) and inhibitory substrates. This is also of considerable interest to understand genetic manipulations designed to engineering higher cellulose content in plants. Of course, genome-scale models represent the vital link between cellular genotype and phenotype. However, genome-scale models contain a static biomass constituting equation. This equation must change in response to genetic and/or environmental perturbations. Previously, it has been shown that a genetic algorithm can draw this link based on carefully crafted objective functions. But, how can predicted phenotype results be tested easily? Cellular composition determination involving mass spectroscopy is time consuming, does not consider spatial variation, and is destructive to dynamically responding cultures. Here, the power of Raman spectroscopy and surface-enhanced Raman spectroscopy (SERS) are demonstrated and interfaced with genome-scale modeling for accurate phenotype predictions. Examples of advanced biofuel tolerance by several bacterial strains and the accumulation of higher cellulose content by Arabidopsis are presented in detail.

See more of this Session: In Silico Systems Biology: Cellular and Organismal Models
See more of this Group/Topical: Topical A: Systems Biology

(534f) Quantitative Model Analysis with Qualitative Biological Data: Stem Cell Autoregulation Mechanisms In the Drosophila Germarium

Michael S. Pargett[1], Robin Harris[2], Hilary Ashe[2] and David M. Umulis[3], (1)Weldon School of Biomedical Engineering, Purdue University, West Lafayette, IN, (2)Faculty of Life Sciences, University of Manchester, Manchester, United Kingdom, (3)Agricultural and Biological Engineering, Purdue University, West Lafayette, IN

Complex organisms maintain stable populations of stem cells to maintain health and support somatic tissues. In the germline stem cells (GSC) of the Drosophila ovary, Bone Morphogenetic Protein (BMP) signaling regulates the decision between stem cell self renewal or differentiation. Key players have been elucidated in the intracellular network that regulates the response to BMPs, however the interaction between intra- and extracellular regulation of BMP signaling remains unknown. Multiple hypotheses on regulatory mechanisms are indistinguishable based on current qualitative comparisons to data. The data available for the germarium are insufficient to optimize model fitness due to a lack of quantitative value in the fluorescent expression and immunohistochemical images. We designed a method to quantitatively constrain the models with existing data, employing multiobjective optimization and optimal scaling of qualitative data. To investigate whether the proposed system is sufficient to autoregulate population, we have developed two mathematical models, considering both intra- and extracellular interactions: a local model for a single cell receiving BMP signaling and a spatial extracellular model of the germarium. We converted the qualitative observation-based data into constraints on a quantitative representation and measured model-data correspondence with an optimal scaling least-squares method. Regulation of a small number of stem cells and the replenishment of an empty niche are investigated by simulation on the spatial model. Autoregulation is reproduced by combining intra- and extracellular regulation in the spatial multi-cell model with a hypothetical mechanism of cell-mediated competition for limiting amounts of ligand.

See more of this Session: In Silico Systems Biology: Cellular and Organismal Models
See more of this Group/Topical: Topical A: Systems Biology

(534g) Dynamic Response of C. Elegans Reproduction to Chronic Heat Stress

Peter B. Winter[1], Patrick D. McMullen[1], Erin Aprison[2], Luis A.N. Amaral[1], Richard I. Morimoto[3] and Ilya Ruvinsky[2], (1)Chemical & Biological Engineering, Northwestern University, Evanston, IL, (2) Ecology and Evolution, University of Chicago, Chicago, IL, (3)Biochemistry, Molecular Biology & Cell Biology, Northwestern University, IL

A major goal of systems biology is to understand how interactions of individual components give rise to emergent properties. The reductionist approach of systems analysis relies on amassing data on each individual component and then uniting the details back into the whole. While thorough, this approach becomes progressively more difficult to implement as the number and complexity of components increases. It is becoming increasingly important to apply a holistic approach to synthesize our understanding into parsimonious models integrating molecular and cellular details into quantifiable phenotypic behavior. The *Caenorhabditis elegans* reproductive system offers an ideal test case for such a model as it has been extensively studied using genetic and biochemical approaches, has a readily quantifiable output, and can easily be probed by exploiting its temperature sensitivity.

We quantify the non-stationary rate of egg laying over time at 6 temperatures using 4,343 individual worms and 191,659 embryos. This data set contains an unprecedented amount of information to train and test our model against the distribution of reproductive outputs. Under ambient conditions, 20°C, brood size distribution for individual worms is normal, whereas prolonged exposure to elevated temperatures can be best described by a combination of two Gaussian distributions. This implies individuals adopt one of two states in their egg laying behavior.

We derive a cellular-scale model using constraints from tissue structure, cellular states, and a minimal set of molecular information to recreate the time-dependent distributions of eggs laid among individual worms. At the model's core are the first principels of process engineering, *mass action kinetics* and *species conservation.* Remarkably the reproductive system bears a striking similarity to production processes engineered by humans. We predict the flow of gametes (oocytes and sperm) through a series of cellular states that signify different stages in development, fertilization, or signaling behavior. Using this framework, it becomes clear that the two distributions in our data set arise due to changes in the upstream development of oocytes. Furthermore, the model suggests a novel protective signaling behavior for stress conditions that prevents waste of the limiting resource, sperm, until the worms return to more favorable conditions.

This model captures the temperature-dependent dynamics of a process that incorporates information over several molecular, cellular and organismal scales. Additionally, we uncover the potential for two novel temperature-dependent switches in two different areas of the reproductive system: the upstream development of oocytes and the regulation of a sperm derived signaling molecule. We argue that top-down mechanistic models are necessary to bridge the gap between scales in biological processes.

See more of this Session: In Silico Systems Biology: Cellular and Organismal Models
See more of this Group/Topical: Topical A: Systems Biology

(590a) Metabolic Network Reconstruction and Genome-Scale Model of Butanol-Producing Strain Clostridium Beijerinckii NCIMB 8052

Caroline B. Milne, Chemical and Biomolecular Engineering, University of Illinois Urbana-Champaign, Urbana, IL and Nathan D. Price, Chemical and Biomolecular Engineering, University of Illinois, Urbana-Champaign, Urbana, IL

Introduction: *In silico* reconstruction of metabolic networks—and subsequent analysis of genome-scale metabolic models—enables a global interrogation of metabolism not possible through standard experiments. Numerous successes have been demonstrated for using these models to guide rational engineering in well-studied microorganisms such as *Escherichia coli* and *Saccharomyces cerevisiae*, and models have been built for over 50 organisms to date. In this work, we use this constraint-based modeling approach to understand *Clostridium beijerinckii* metabolism, and to help guide rational engineering experiments to improve production of butanol—an important chemical and potential fuel additive or replacement. *C. beijerinckii* is an attractive organism for strain design geared towards the production of butanol because it (i) naturally produces the highest recorded butanol concentrations as a byproduct of fermentation; and (ii) can co-ferment pentose and hexose sugars (the primary sugars in lignocellulosic hydrolysate). Interrogating the metabolism of this microorganism from a systems viewpoint will allow us to simulate the global effect of various genetic modifications.

Materials and Methods: The genome-scale metabolic network for *C. beijerinckii* was reconstructed by merging annotation data specific to this microorganism from three major databases: KEGG, The SEED and BioCyc. The network was further refined using both computational algorithms (GapFind, GapFill) and manual curation to minimize the number of gaps, or unusable reactions. A series of balances (e.g. mass, energy) and bounds (e.g., flux capacities, thermodynamics) were applied to a matrix representation of the metabolic network to establish a computable genome-scale model. We then used Flux Balance Analysis to determine the distribution of reaction fluxes that optimize our defined objective (biomass). To assess the predictive accuracy of the model, we conducted batch fermentation experiments to determine substrate uptake and byproduct secretion rates (corresponding to model inputs and outputs, respectively). We also conducted metabolomics profiling under a variety of conditions to further complete the model and assess its predictability. Finally, the network was probed to identify and understand reactions having the largest impact on butanol formation.

Results and Discussion: We present the first genome-scale metabolic model (*i*CM926) for *C. beijerinckii,* containing 926 genes, 939 reactions, and 881 metabolites. The semi-automated procedure we employed to build the model provided a rapid method for bringing the genome of a lesser-characterized microorganism to life. Specifically, this procedure capitalized on annotations available in multiple genome annotation databases (KEGG, BioCyc, The SEED) to maximize the scope of the network, and utilized computational algorithms to fill network gaps and improve model completeness. Interestingly, we found that the overlap in information between the three databases was only 34%; such a small agreement highlights the importance of exercising caution when building genome-scale models, which are often based upon one primary annotation source. To ensure predictive accuracy of our *i*CM926 model and assess the contribution of each database, we evaluated the ability of the model to reproduce experimentally observed substrate uptake and product production rates, and examined the source database of each reaction in the resulting flux distribution. The *i*CM926 model, when simulated with an optimal growth objective and fixed uptake rates for only acetate and glucose, was unable to reproduce measured production rates of butyrate, acetone, butanol and ethanol. However, with additional constraints, the experimental conditions could be more accurately simulated. Notably, a significant number of actively utilized reactions in the latter simulations originated from the set of reactions that overlapped between all three databases ($P = 3.52\times10^{-9}$, Fisher's exact test). In our simulations, inhibition of the hydrogenase reaction was found to have the largest effect on butanol formation—a relationship that has been experimentally observed as well.

Conclusions: *In silico* modeling of genome-scale networks offers a promising method for globally interrogating the metabolism of *C. beijerinckii*, a microorganism of industrial interest for butanol production. The *i*CM926 model is a predictive model that accurately simulates physiological behavior and provides insight into the underlying mechanisms of microbial butanol production. As such, the model will be instrumental in efforts to better understand, and metabolically engineer, this microorganism for improved butanol production. Additionally, the construction of *i*CM926 for a lesser-characterized microorganism highlighted important areas of investigation (e.g. genome annotations, objective functions) for model-building efforts going forward.

See more of this Session: In Silico Systems Biology: Cellular and Organismal Models II
See more of this Group/Topical: Topical A: Systems Biology

(590b) A Novel FBA Strategy for Metabolic Modeling of Cyanobacteria

Tirthankar Sengupta, Mani Bhushan and Pramod P. Wangikar, Chemical Engineering Department, Indian Institute of Technology Bombay, Powai, Mumbai, India

Metabolic modeling refers to modeling the gamut of metabolic reactions occurring in an organism. Flux Balance Analysis (FBA) is a constraint based metabolic modeling strategy. The main premise in FBA is that an organism is designed by nature to maximize its' chances of survival. In quantitative terms this translates to the hypothesis that an organism tries to maximize its growth rate through the metabolic reactions occurring within the organism. Thus in FBA, the biomass flux (the flux of the reaction describing the formation of biomass from precursor metabolites) is the objective function that needs to be maximized. The constraints in the model arise from mass balances on the metabolites. While internal metabolites are assumed to be at steady state, external metabolites (nutrients or metabolites excreted by an organism) can be at unsteady state. The steady state assumption stems from the hypothesis of an organism trying to maximize its' growth rate. In such a scenario, all absorbed nutrients must finally end up as biomass and hence the initial and final amounts of internal metabolites must remain the same. Thus considering the overall behavior, internal metabolites must remain at steady state. In the FBA model, all fluxes are assumed to be unbounded. This assumption would hold true when enough amounts of enzymes are present in an organism at all times to catalyze any amount of substrate. Since enzyme levels in an organism vary in response to extracellular stimuli, leaving the flux of a reaction unconstrained would not be a proper assumption. In this work, a FBA model is proposed which accounts for the fact that enzyme levels within cells vary with perturbations in the environment to which the cells are exposed to. In the proposed model the fluxes of reactions are not left unconstrained. Instead, the flux of a reaction is constrained depending on the amount of the corresponding enzyme. Enzyme levels are assumed to be directly proportional to the corresponding mRNA levels and information about the latter is obtained from microarray experimental data. Technique presented by Sengupta et al. (2010) is used to interpret the microarray data. This technique is adapted from a normalization strategy proposed by Chen et al. (1997).

Cyanobacteria are photosynthetic prokaryotes. It is believed that they are one the earliest forms of bacteria to have evolved on earth. It is also believed that they were responsible for introducing oxygen into the earth's atmosphere. Further, cyanobacteria produce useful chemicals such as hydrogen (a potential alternate fuel) by virtue of their metabolism. Cyanobacteria thus form a very important species of bacteria to be studied both from a fundamental as well as a commercial viewpoint. In this work, we modeled the metabolism of Synechocystis species strain PCC 6803 using the proposed FBA modeling strategy. The metabolic model was borrowed from Shastri and Morgan (2005). Microarray experimental data of the organism was obtained from the KEGG database which contains gene expression data of the organism under a variety of environmental conditions. Several metabolic topologies were obtained which were different from the topology obtained using the usual FBA modeling strategy. For example, one topology was obtained in which the pentose phosphate pathway became active but was not active in the metabolic map obtained from the usual FBA. The proposed FBA strategy thus gives additional insights into the metabolism of an organism that cannot be realized from a usual FBA.

References

1. Y. Chen, E. R. Dougherty and M. L. Bittner, "Ratio-based decisions and the quantitative analysis of cDNA microarray images," *Journal of Biomedical Optics* **2**(4), 364-374 (1997).
2. Shastri A. A. and Morgan J. A., Flux balance analysis of photoautotrophic metabolism, Biotechnol. Prog., 2005, 21, 1617-1626.
3. T. Sengupta, M. Bhushan and P. P. Wangikar, Annual Progress Seminar Report, 2010, Department of Chemical Engineering, Indian Institute of Technology Bombay.

See more of this Session: In Silico Systems Biology: Cellular and Organismal Models II
See more of this Group/Topical: Topical A: Systems Biology

(590c) Constraint-Based Modeling of Cyanobacterial Metabolism

Trang T. Vu[1], Sergey M. Stolyar[2], Grigoriy E. Pinchuk[2], Allan E. Konopka[2], Alex S. Beliaev[2] and Jennifer L. Reed[1], (1)Department of Chemical and Biological Engineering, University of Wisconsin - Madison, Madison, WI, (2)Pacific Northwest National Laboratory, Richland, WA

The U.S. currently needs alternative renewable sources of energy to reduce our dependence on fossil-fuels, which are becoming increasingly limited and expensive. Research investments and efforts focused on developing clean and efficient processes for sustainable fuel production, of which hydrogen is considered as one of the promising future fuels [1]. Among bio-based hydrogen-producing processes, oxygenic photosynthesis receives great interest, since light energy is converted to chemical energy using water as electron donor for reduction of electron carriers (ferredoxin and NADPH), which can subsequently be used to generate hydrogen. Oxygenic photosynthetic microorganisms display varying levels of light conversion efficiencies, which ultimately translate into different rates of electron transfer, growth, and hydrogen production. Developing a model for such organisms will offer solutions to fundamental science questions of how different factors are interconnected at a system level, and affect cellular phenotypes. In addition, the model will have the ability to serve as an *in silico* tool for manipulating photosynthetic microorganisms to act as catalysts for solar energy conversion and will potentially allow development of a highly efficient biofuel production process.

We have built a genome-scale constraint-based metabolic model for *Cyanothece* sp. ATCC 51142, a unicellular diazotrophic cyanobacterium that can temporally separate the process of light-dependent autotrophic growth and glycogen accumulation from N_2 fixation. The resulting model currently includes 806 genes, 585 metabolites, and 664 reactions accounting for common pathways such as central metabolism, respiration, nucleotide and amino acid biosynthesis, and those that are more unique to cyanobacteria such as photosynthesis, carbon fixation, and cyanophycin production. Photosynthesis was modeled as a set of three sequential reactions that involve photosystem II, cytochrome b_6f complex, and photosystem I. This was done in order to study how light intensities of different wavelengths of light and separate photosystem activities affect electron flow through photosynthesis and respiration, nitrogen source utilization, cellular growth and hydrogen production.

Data on glycogen requirement for nitrogen fixation in the dark [2] and amino acid utilization as sole nitrogen sources have been used to test and refine the model. Additionally, we have used the model to estimate energy requirements for *Cyanothece* sp. ATCC 51142 using experimental batch data collected with different light wavelengths of various intensities. We have subsequently used the model to investigate how flux through different photosystems affects cellular growth and fluxes through linear and cyclic photosynthesis pathways, cytochrome oxidases, NADH dehydrogenases, Mehler reactions, and ferredoxin-NADP reductase.

Using a custom-built photobioreactor, which allows for the control and monitoring of incident and transmitted light, we have measured biomass composition, gene expression and protein expression of *Cyanothece* sp. ATCC 51142 under nitrogen- and light-limited chemostat conditions. These experimental datasets have been incorporated into the model to improve flux predictions for both chemostat conditions. In the future, the model will be used to simulate hydrogen production rates, and to suggest metabolic engineering strategies for improving hydrogen and chemical production.

References

1. Dunn S: **Hydrogen futures: toward a sustainable energy system**. *International Journal of Hydrogen Energy* 2002, **27**(3):235-264.

2. Schneegurt MA, Sherman DM, Sherman LA: **Growth, physiology and ultrastructure of a diazotrophic cyanobacterium, Cyanothece sp. strain ATCC 51142 in mixotrophic and chemoheterotrophic cultures**. *Journal of Phycology* 1997, **33**:632-642.

See more of this Session: In Silico Systems Biology: Cellular and Organismal Models II
See more of this Group/Topical: Topical A: Systems Biology

(590d) Metagenome Based Metabolic Network Analysis of Human Microbiome

Yu Chen, Department of Chemical Engineering, University of Michigan, Ann Arbor, MI and Xiaoxia Nina Lin, Chemical Engineering, University of Michigan - Ann Arbor, Ann Arbor, MI

The human body harbors various microbial communities (i.e. human microbiome) that are believed to affect the host health but remain largely enigmatic. Recent years have seen rapidly increasing use of metagenomic sequencing for study of natural microbial communities and the human microbiome has become one of the latest targets of such metagenomic investigation. In this work, we set out to make use of these large-scale metagenomic datasets to elucidate the metabolic functions of the human microbiome and the microbe-microbe-host interactions, by developing and utilizing a metabolic network modeling framework.

Using a gut microbiota metagenomic dataset generated from sampling of 124 European individuals, we reconstructed *in silico* metabolic networks of various phylogenetic groups at different taxonomic levels. To represent variations among samples (people), the frequency (discovery rate) of each metabolic reaction was calculated prior to network reconstruction. Then minimal metabolic networks for biomass synthesis were generated for different phylogenetic groups, each consisting of 1,300-2,200 reactions. Furthermore, minimal reaction sets for the synthesis of individual biomass components were examined based on the entire dataset and their discovery rates in different phylogenetic groups were compared. Our results suggested distribution of certain metabolic functions, such as biosynthesis of short-chain fatty acids and vitamins, in the microbial community.

To directly analyze the interactions among microbiome members, we also developed multiple-compartment network models that consider the most abundant groups at the same time. It was found that many metabolic interactions potentially exist among the microbial groups, e.g. between *Bacteroidetes*, *Firmicutes* and *Proteobacteria*. These results indicated that metabolic cooperation among microbes is common and might be important for sustaining robust human microbiota.

See more of this Session: In Silico Systems Biology: Cellular and Organismal Models II
See more of this Group/Topical: Topical A: Systems Biology

(590e) In Silico Metabolic Model of Rhodococcus Erythropolis to Study and Improve Desulfurization

Shilpi Aggarwal, Chemical and Biomolecular Engineering, National University of Singapore, Singapore, Singapore, Iftekar A. Karimi, National University of Singapore, Singapore, Singapore and Dong-Yup Lee, Chemical and Biomolecular Engineerig, National University of Singapore, Singapore, Singapore

In silico Metabolic Model of *Rhodococcus erythropolis* to Study and Improve Desulfurization

Shilpi Aggarwal[1], I A Karimi[1], Dong-Yup Lee[1,2]

[1]Dept of Chemical & Biomolecular Engineering, National University of Singapore, Singapore 117576

[2]Bioprocessing Technology Institute, 20 Biopolis Way, #06-01, Centros, Singapore 138668

The increasingly stringent rules for ultra-low-sulfur fuels have inspired efforts to improve existing desulfurization techniques and develop new, efficient, and more economical methods. Hydrodesulfurization, the prevalent method for desulfurization, is a chemical process that is energy-intensive, expensive, and incapable of desulfurizing certain recalcitrant sulfur compounds such as benzothiophene, dibenzothiophene (DBT), and their derivatives present in fossil fuels [1]. Studies aimed at evolving new methods of desulfurization have recognized biodesulfurization as a potential alternative.

Biodesulfurization involves the use of either enzymes or whole cells for reducing sulfur content of the fossil fuels. It is advantageous, as it can proceed under ambient conditions, does not involve the loss of calorific value [2], and is relatively economical. *Rhodococcus erythropolis* IGTS8 was isolated as the first microorganism that can specifically desulfurize polyaromatic sulfur heterocycles (PASHs) such as DBT and its derivatives via the 4S pathway. Over the last two decades, several strains of *Rhodococcus* have been studied for their desulfurization ability [2]. The desulfurization rates obtained in a biodesulfurization process using naturally occurring bacterial cultures are too low for commercialization [3]. Despite numerous efforts for increased specific desulfurization activity with various manipulations at the genetic level, desirable desulfurization rates are yet to be attained.

The complex interactions among the various metabolic pathways and associated reactions largely determine the metabolic fluxes [4] within an organism and hence its cellular activities and phenotypes (here, desulfurization activity). However, most studies have targeted the 4S pathway exclusively, and a holistic systems study of various intracellular activities has not been reported. Therefore, it is critical to study the desulfurizing characteristics of *R. erythropolis* by considering its interactions and dependence on the other parts of its metabolism.

This work represents the very first stoichiometric model of *R. erythropolis* to study its intracellular metabolic processing of various sulfur compounds. It consists of the sulfur metabolic pathway along with the other biochemical pathways such as central metabolism, amino acids biosynthetic pathways, etc. It describes the dependence of the flux through sulfur metabolism on factors such as the supply of reducing equivalents and demand for sulfur containing precursor metabolites. It has been successfully validated using the experimental data available in literature. The model predicts biomass growth close to the experimental values with the measured DBT uptake rates [5] used as inputs. Moreover, it shows the effects of alternate sulfur sources (i.e., sulfate and DBT) on the growth rates which are in close agreement with the experimental observations [6, 7]. Based on the model findings, we have been able to propose an alternate hypothesis for the effect of sulfate on desulfurizing activity. Successful predictions on the suitability of carbon sources [8] by the model have led us to determine the effectiveness of additional carbon sources and other medium components for achieving higher *in silico* desulfurizing activity and growth of this biocatalyst. The determination and comparison of the fluxes through other host functions provide insights into the relative influence of the various enzymatic activities on the extent of desulfurization exhibited by *R. erythropolis*. The analysis of the metabolic network has also enabled us to identify the various metabolic engineering approaches such as gene(s) knockouts, over expression of certain gene(s), etc. that may be instrumental in enhancing the desulfurization activity of *R. erythropolis* effectively.

1. Song, C., *An overview of new approaches to deep desulfurization for ultra-clean gasoline, diesel fuel and jet fuel.* Catalysis Today, 2003. **86**(1-4): p. 211-263.

2. Soleimani, M., A. Bassi, and A. Margaritis, *Biodesulfurization of refractory organic sulfur compounds in fossil fuels.* Biotechnology Advances, 2007. **25**(6): p. 570-596.

3. Kilbane II, J.J., *Microbial biocatalyst developments to upgrade fossil fuels.* Current Opinion in Biotechnology, 2006. **17**(3): p. 305-314.

4. Raman, K. and N. Chandra, *Flux balance analysis of biological systems: Applications and challenges.* Briefings in Bioinformatics, 2009. **10**(4): p. 435-449.

5. Ohshiro, T., Y. Hine, and Y. Izumi, *Enzymatic desulfurization of dibenzothiophene by a cell-free system of Rhodococcus erythropolis D-1.* FEMS Microbiology Letters, 1994. **118**(3): p. 341-344.

6. Omori, T., et al., *Desulfurization of alkyl and aromatic sulfides and sulfonates by dibenzothiophene-desulfurizing Rhodococcus sp. strain SY1.* Bioscience, Biotechnology and Biochemistry, 1995. **59**(7): p. 1195-1198.

7. Honda, H., et al., *High cell density culture of rhodococcus rhodochrous by pH-stat feeding and*

dibenzothiophene degradation. Journal of Fermentation and Bioengineering, 1998. **85**(3): p. 334-338.

8. Yan, H., et al., *Increase in desulfurization activity of Rhodococcus erythropolis KA2-5-1 using ethanol feeding.* Journal of Bioscience and Bioengineering, 2000. **89**(4): p. 361-366.

See more of this Session: In Silico Systems Biology: Cellular and Organismal Models II
See more of this Group/Topical: Topical A: Systems Biology

(590f) Contrasting the Metabolic Capabilities of Multiple Cyanobacterial Species

Rajib Saha[1], Bertram M. Berla[2], Thanura Elvitigala[3], Lawrence E Page[3], Himadri B Pakrasi[3] and Costas D. Maranas[1], (1)Department of Chemical Engineering, The Pennsylvania State University, University Park, PA, (2)Dept. of Energy, Environmental, and Chemical Engineering, Washington University in St. Louis, St. Louis, MO, (3)Dept. of Biology, Washington University in St. Louis, St. Louis, MO

Cyanobacteria are an important group of photoautotrophic organisms that can synthesize valuable bioproducts by harnessing solar energy. Cyanobacteria contribute significantly to biological carbon sequestration, O_2 production and the nitrogen cycle. They exhibit robust growth under diverse environmental conditions and have minimal nutritional requirements. They are also endowed with high photosynthetic efficiencies and diverse metabolic capabilities that confer the ability to convert solar energy into a variety of biofuels and their precursors. However, less well studied are the similarities and differences that exist between the metabolic capabilities of different species of cyanobacteria that may contribute toward their niche biological functions. In this talk, we present progress in the development of genome-scale models of two phylogenetically close cyanobacterial species, namely *Cyanothece 51142* (*i*RS703) and *Synechocystis 6803* (*i*RS 661) to address these issues. Model *i*RS703 is comprised of 703 genes, 1654 reactions and 1837 metabolites; whereas model *i*RS661 spans 661 genes, 1693 reactions and 1896 metabolites. As many as 1484 reactions and 1705 metabolites are common in these two models. All reactions are elementally and charged balanced and localized into three different intracellular compartments (i.e., cytoplasm, carboxysome and thylakoid lumen). GPR (gene-protein-reaction) associations are also established based on the functional annotation information and homology prediction and categorized into five major classes of monofunctional proteins, multifunctional proteins, isozymes, multimeric proteins and protein complexes. We describe results from performing flux balance analysis under different physiological conditions, (i.e., phototrophic, chemotrophic and mixotrophic) and explore their impact on metabolic capabilities such as biomass formation, nitrogen fixation and hydrogen yield of these two cyanobacterial species.

See more of this Session: In Silico Systems Biology: Cellular and Organismal Models II
See more of this Group/Topical: Topical A: Systems Biology

(590g) Genome Scale Analysis and Dynamic Study of Methanogen, Methanosarcina Barkeri

Matthew C. Gonnerman, Chbe, UIUC, Urbana, IL and Nathan D. Price, Chemical and Biomolecular Engineering, University of Illinois, Urbana-Champaign, Urbana, IL

M. barkeri is a methanogen, meaning that methane is the reduced product for energy generation. *M. barkeri* can grow on several substrates including methanol, acetate, methyl amines and a combination of hydrogen and carbon dioxide. The *Methanosarcina barkeri* genome was published in 2006 after the metabolic reconstruction of *M. barkeri*, iAF692, was published. The iAF692 reconstruction preformed the first large scale simulations of an Archaea. Since the publication of iAF692, our deeper understanding of Archaea biology has shed light on previously unclear metabolic pathways. Reactions in methanogenesis, cofactor biosynthesis, pentose phosphate pathway and others have been updated to reflect Archaea biosynthesis pathways. This improved insight has led to an increase in predictive accuracy for knockouts, dynamic simulations and growth yield predictions for the updated *M. barkeri* reconstruction, iMG746. As with other metabolic reconstructions, iMG746 will be a useful tool in guiding experimental studies and predicting cellular behavior on a genome scale.

See more of this Session: In Silico Systems Biology: Cellular and Organismal Models II
See more of this Group/Topical: Topical A: Systems Biology

(653a) Predicting Evolution

Pornkamol Unrean, Chemical Engineering and Materials Science, and Biotechnology Institute, University of Minneasota, Minneapolis/St. Paul, MN, Pedro V. Pena, Chemical Engineering and Materials Science, and BioTechnology Institute, University of Minnesota, Minneapolis/St. Paul, MN and **Friedrich Srienc**, Chemical Engineering and Materials Science, and BioTechnology Institute, University of Minnesota, Minneapolis/St Paul, MN

The metabolic network of a cell can be decomposed into discrete elementary modes that contribute, each with a certain probability, to the overall flux through the metabolism. These modes are cell function supporting, fundamental pathways that represent permissible, discrete states of the metabolism. Recently, we have theoretically shown that the flux through individual elementary modes is distributed according to the Boltzmann distribution law if cells developed regulation patterns in an unbiased way. Adaptive evolution experiments suggest that cells indeed evolve towards such state. Moreover, knowledge of this state enables prediction of the adaptive mutations occurring during evolution. A summary of the developed theory will be presented together with supporting experimental evidence.

See more of this Session: In Silico Systems Biology: Intracellular Signaling and Gene Regulation
See more of this Group/Topical: Topical A: Systems Biology

(653b) The Antibiotic Resistance Game: The Genetic Regulation of Antibiotic Resistance Transfer In Enterococcus Faecalis

Anushree Chatterjee, Chemical Engineering and Materials Science, University of Minnesota, Minneapolis, MN, Che-Chi Shu, Purdue University, West Lafayette, IN, Yiannis Kaznessis, Department of Chemical Engineering and Materials Science, University of Minnesota, Minneapolis, MN, Doraiswami Ramkrishna, School of Chemical Engineering, Purdue University, West Lafayette, IN, Gary M. Dunny, Department of Microbiology, University of Minnesota, Minneapolis, MN and Wei-Shou Hu, Chemical Engineering and Materials Science, University of Minnesota, Minneapolis

One of the main factors driving the rise of drug resistant microbes is the transfer of antibiotic resistance genes present on mobile plasmids between donor and recipient cells via the mechanism of conjugation. In order to combat microbial drug resistance, novel strategies need to be developed to block such transmission of antibiotic resistance. In this work, we study the gene regulatory components involved in transfer of tetracycline resistance conferring plasmid pCF10 between plasmid-carrying donor cells and plasmid-deficient recipient cells in bacterium *Enterococcus faecalis*. The onset of conjugation is tightly regulated via multi-layered regulation offered by two-key genes *prgQ-prgX* present on pCF10 in response to a signaling molecule secreted by recipient cells. Using mathematical modeling and experimentation, we describe a novel mechanism of gene-regulation due to transcriptional interference and interaction between sense-antisense transcripts in the *prgX/prgQ* operon, which causes the system to demonstrate a bistable genetic switch (Chatterjee *et al.*, 2011). Once the genetic-switch is ON, expression of approximately 30 genes downstream of *prgQ* gene, encoding various proteins involved in transfer of the plasmid, are induced. Induction also results in production of an inhibitor of conjugation, giving rise to negative feedback loop. Mathematical model and experiments indicate that such a negative feed-back loop causes downstream genes to demonstrate a "pulse" of expression, with nearly 2-3 orders of magnitude increase and decrease in expression within a short duration post-induction. Furthermore, proteomic analysis also shows dynamic expression of proteins encoded by the downstream genes. An interplay of positive and negative feedback loops allows the system to quickly transition between ON and OFF states and is critical both for the transfer of plasmid and survival of the donor cell. Studying both the turning-ON and turning-OFF mechanisms of the switch allows identification of potential drug targets for blocking transmission of antibiotic resistance for use in future therapy.

Chatterjee, A., Johnson, C.M., Shu, C.-C., Kaznessis, Y.N., Dunny, G.M., and Hu, W.-S. (2011) Convergent transcription confers a Bistable switch in Enterococcus faecalis conjugation. *Proc Natl Acad Sci U S A (in press).*

See more of this Session: In Silico Systems Biology: Intracellular Signaling and Gene Regulation
See more of this Group/Topical: Topical A: Systems Biology

(653c) A Novel Multi-Layer Inference Approach to Reconstruct Conditional-Specific Gene Network

Ming Wu, Computer Science and Engineering, Michigan State University, East Lansing, MI and Christina Chan, Michigan State Uiversity, East Lansing, MI

The recent advent of high-throughput microarray data has enabled the global analysis of the transcriptome, driving the development and application of computational approaches to study gene regulation on the genome scale, by reconstructing *in silico* the regulatory interactions of the gene network. Our approach differs from most of the currently available approaches by inferring transcriptional and post-transcriptional regulation. In contrast to approaches that infer both types of regulation, our methodology captures biological features at different layers rather than use a black box approach to "learning" the interaction. Furthermore, our framework does not require perturbation on each gene or time-series data to uncover the regulatory interactions. Thus far, the development and application of the theory and tools for network inference have been predominantly based on statistical dependencies between gene expressions. These methods have been shown to capture only part of the relevant biological information. These current methods generate many candidates/hypotheses when applied to real biological systems, making it difficult to experimentally assess each one. Furthermore a critical analysis of the underlying assumptions of most *in silico* learning approaches applied to expression data [1] demonstrate that statistical dependencies alone do not readily provide direct mechanistic insights or interactions from functional genomics. We aim to address these challenges and propose a novel multi-layer network inference approach which integrates gene expression data and transcriptional regulation information to recover a specific gene network for a given condition. Our reverse engineering framework consists of three layers. The first layer integrates microarray data from a diverse set of conditions to provide a common context of gene behaviors and then identifies the most specific candidate genes for a given condition. In the second layer we apply a state-of-art causal filtering method that combines information on the transcriptional regulatory network to reconstruct the regulatory pathway to the candidate genes. In the third layer we develop a novel method of inferring transcription factor activity to identify the important regulators in the gene network that can account for post-transcriptional regulation. By integrating multiple layers of learning, our framework captures different biological features in the transcriptional regulation to achieve an accurate reconstruction of condition-specific gene networks. We establish the accuracy of our methodology against a synthetic dataset as well as the yeast dataset. Finally we extend the framework to the application of human diseases, such as cancer.

[1] Wu, M and Chan, C. (2011) Learning transcriptional regulation on a genome scale: a theoretical analysis based on gene expression data, **Briefings in Bioinformatics**, accepted

See more of this Session: In Silico Systems Biology: Intracellular Signaling and Gene Regulation
See more of this Group/Topical: Topical A: Systems Biology

Regularization of Ill-conditioned Inverse Problems Resulting from Determining Transcription Factor Profiles from Fluorescent Reporter Systems

Loveleena Bansal, Yunfei Chu, Carl Laird, and Juergen Hahn

Artie McFerrin Department of Chemical Engineering

Texas A&M University

A quantitative description of transcription factor dynamics can aid in understanding the response of cells to external stimuli [1, 2] as they initiate the transcription and translation process in the cell. A number of researchers in the last two decades have used fluorescence based reporter systems for continuous and non-invasive monitoring of gene expression and transcriptional activity [3, 4]. The fluorescence of proteins, such as the green fluorescent protein (GFP), observed from fluorescence microscopy or a fluorescent plate reader, can be used as an indicator of the activation of transcription factors. A number of mathematical models describing transcription, translation and fluorophore formation of fluorescent proteins have been developed [5, 6, 7, 8, 9]. These models have been used for estimating the mRNA or transcription dynamics from gene expression data. In most of these works, a certain nature of the dynamic profile of a compound was assumed and then the parameters were estimated to characterize the profile. However, this restricts the shapes that the estimated profiles can take. Thus, the main aim of this study is to develop an inverse problem formulation such that any shape of dynamic transcription factor profiles can be estimated from the fluorescent intensity profiles of fluorescent reporter systems.

The ODE model developed by [10] which describes transcription, translation and fluorophore formation of GFP has been used in this work. The input to this ODE model is the time-dependent concentration of transcription factor while the observed fluorescence is treated as the output. It is assumed that the profile for transcription factor is piecewise constant over a discretization interval and it only changes between the discretization points. Thus, the ODE model describing transcription/ translation is discretized and presented in the form of linear algebraic equations. The resulting discrete inverse problem is also highly ill-conditioned. Two regularization methods - truncated singular value decomposition (TSVD) and Tikhonov regularization - have been used for the solution of the inverse problem at hand and the results from the two methods are compared. The method of choosing the regularization parameters for each technique is discussed is detail.

The regularization techniques have been applied to simulated data and experimental data for the transcription factor STAT3 in hepatocytes which are continuously stimulated by IL-6. For simulated data, both methods performed equally well if little noise was present. However, if the noise levels were large, the profiles estimated by Tikhonov regularization were a slightly better representation of the actual dynamics. For the case study involving real data, the dynamic profiles of the transcription factor STAT3 obtained after solving the inverse problem were in close agreement with results reported in the literature. The estimated STAT3 profile is shown in Figure 1. STAT3 dynamics was found to be oscillatory in nature with a large initial peak followed by a smaller peak at around 5-6 hours and potentially another peak at approximately 10 hours. These results are consistent with Western blot data [11] as well as simulation results of the IL-6 signal transduction pathway [12] given in the literature. Similar to the investigations based upon simulated data, the results returned by the Tikhonov regularization seem to provide a better representation of the STAT3 dynamics as compared to TSVD.

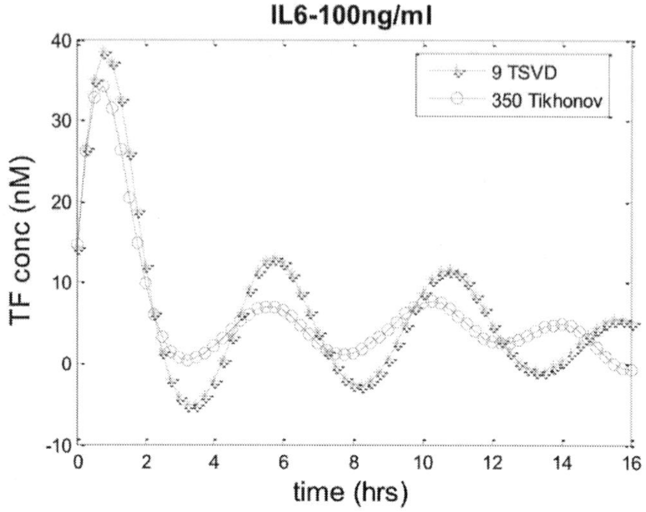

Figure 1 Estimated STAT3 profile using TSVD and Tikhonov Regularization with a regularization parameter of 9 and 350 respectively

References

[1] Luscombe, N. M., Babu, M. M., Yu, H., Snyder, M., Teichmann, S. A. and Gerstein, M. (2004). Genomic analysis of regulatory network dynamics reveals large topological changes. *Nature,* Vol. 431, No. pp. 308-312

[2] Jothi, R., Balaji, S., Wuster, A., Grochow, J. A., Gsponer, J., Przytycka, T. M., Aravind, L. and Babu, M. M. (2009). Genomic analysis reveals a tight link between transcription factor dynamics and regulatory network architecture. *Molecular Systems Biology* Vol. 5, No. pp.

[3] Chalfie, M., Tu, Y., Euskirchen, G., Ward, W. and Prasher, D. (1994). Green fluorescent protein as a marker for gene expression. *Science,* Vol. 263, No. 5148, pp. 802-805

[4] van Roessel, P. and Brand, A. H. (2002). Imaging into the future: visualizing gene expression and protein interactions with fluorescent proteins. *Nat Cell Biol,* Vol. 4, No. 1, pp. E15-E20

[5] Subramanian, S. and Srienc, F. (1996). Quantitaive Analysis of transient gene expression in mammalian cells using the green fluorescent protein. *Journal of Biotechnology,* Vol. 49, No. pp. 137-151

[6] Leveau, J. H. J. and Lindow, S. E. (2001). Predictive and Interpretive Simulation of Green Fluorescent Protein Expression in Reporter Bacteria. *Journal of Bacteriology,* Vol. 183, No. 23, pp. 6752-6762

[7] Finkenstädt, B., Heron, E. A., Komorowski, M., Edwards, K., Tang, S., Harper, C. V., Davis, J. R. E., White, M. R. H., Millar, A. J. and Rand, D. A. (2008). Reconstruction of transcriptional dynamics from gene reporter data using differential equations. *Bioinformatics,* Vol. 24, No. 24, pp. 2901-2907

[8] Wang, X., Errede, B. and Elston, T. C. (2008). Mathematical Analysis and Quantification of Fluorescent Proteins as Transcriptional Reporters. *Biophysical Journal,* Vol. 94, No. 6, pp. 2017-2026

[9] Jong, H. d., Ranquet, C., Ropers, D., Pinel, C. and Geiselmann, J. (2010). Experimental and computational validation of models of fluorescent and luminescent reporter genes in bacteria. *BMC Systems Biology,* Vol. 4, No. 55, pp.

[10] Huang, Z., Senocak, F., Jayaraman, A. and Hahn, J. (2008). Integrated modeling and experimental approach for determining transcription factor profiles from fluorescent reporter data. *BMC Systems Biology,* Vol. 2, No. 1, pp. 64

[11] Fischer, P., Lehmann, U., Sobota, R. M., Schmitz, J., Niemand, C., Linnemann, S., Haan, S., Behrmann, I., Yoshimura, A., Johnston, J. A., Muller-Newen, G., Heinrich, P. C. and Schaper, F. (2004). The role of the inhibitors of interleukin-6 signal transduction SHP2 and SOCS3 for desensitization of interleukin-6 signalling. *Biochemistry Journal,* Vol. 378, No. 2, pp. 449 - 460

[12] Singh, A., Jayaraman, A. and Hahn, J. (2006). Modeling Regulatory Mechanisms in IL-6 Signal Transduction in Hepatocytes. *Biotechnology and Bioengineering,* Vol. 95, No. 5, pp. 850 - 862

(653e) Systems Modeling of Epidermal Growth Factor and Nerve Growth Factor Signaling: New Insights Into Transient Vs. Sustained ERK Activation Controlling Cell Fate Decisions

Marc R. Birtwistle, Alexander von Kriegsheim and Walter Kolch, Systems Biology, University College Dublin, Belfield, Ireland

Transient vs. sustained ERK activation in PC12 cells is a paradigm for how epidermal growth factor (EGF) and nerve growth factor (NGF) elicit distinct proliferation vs. differentiation cell fate decisions. Despite much effort to uncover how transient vs. sustained ERK dynamics arise, detailed mechanisms remain unclear. To understand transient vs. sustained ERK signaling mechanistically, we have constructed an ordinary differential equation model of EGF and NGF signaling to ERK in PC12 cells that is trained, and thereby consistent with quantitative data from a variety of literature sources and our own experimental data. In particular, the basis of the model topology is largely derived from a quantitative proteomics screen for ERK interaction partners where we surveyed differential dynamic changes in ERK interacting proteins in PC12 cells treated with EGF or NGF. The emergent model structure suggests that only cell surface receptors can stimulate the PLCγ/PKC pathway, activation of which inhibits the key ERK cascade inhibitor RKIP and also activates Ras through the GRP family of GEFs. We furthermore uncover a putative novel positive feedback loop from ERK to NF1, the predominant RasGAP in PC12 cells. The main model prediction is that rapid EGF-induced internalization of the EGF receptor and weak PLCγ activation, as opposed to NGF-induced surface retention of the TrkA receptor and strong PLCγ activation, generates the distinct ERK activation kinetics. Validation experiments show that increasing plasma membrane levels of active EGF receptor by stimulating with TGFα sustains the ERK response relative to that of EGF, whereas PLCγ knockdown makes the NGF-induced ERK response transient. We conclude that receptor compartmentalization and the PLCγ pathway play an important role in determining transient vs. sustained ERK signaling.

See more of this Session: In Silico Systems Biology: Intracellular Signaling and Gene Regulation
See more of this Group/Topical: Topical A: Systems Biology

(653f) A Novel Mathematical Model of TGF-β-Induced Apoptosis Signaling Pathways

Seung-Wook Chung[1], Carlton R. Cooper[2], Mary C. Farach-Carson[3] and Babatunde A. Ogunnaike[1], (1) Chemical Engineering, University of Delaware, Newark, DE, (2)Biological Sciences, University of Delaware, Newark, DE, (3)Biochemistry and Cell Biology, Rice University, Houston, TX

Transforming growth factor β (TGF-β) signaling plays a vital role in regulating a wide range of cellular processes in both healthy and cancerous cells. In normal or pre-malignant epithelia, TGF-β is well known to inhibit cell proliferation and to induce programmed cell death (or apoptosis). However, in late stages of cancer progression, the role of TGF-β appears to shift in favor of tumor promotion, apparently facilitating epithelial-to-mesenchymal transition (EMT), motility and invasion. While the full details of the mechanisms underlying how TGF-β induces such diverse and contradictory responses during cancer progression are not yet completely elucidated, recent computational studies of the canonical SMAD-mediated signaling pathway have produced plausible explanations that have significantly improved our understanding of these mechanisms [1]. Still, many aspects of TGF-β's role in healthy and cancerous cells remain completely unexplored, including how TGF-β elicits pro-apoptotic effects and how such effects are markedly weakened in cancer cells. A coherent quantitative understanding of the mechanism of TGF-β-induced apoptosis, an indispensable component of any potentially successful TGF-β-mediated targeted cancer treatment, is completely unavailable at present.

In this study, we present for the first time a mathematical modeling framework for understanding the complex signaling behavior of TGF-β-induced apoptosis. The model incorporates various TGF-β branch pathways, including the TAK1-MKK3/6-p38, Smad-GADD45β-MTK1, and putative ARTS pathways, which are coupled with downstream caspase activation cascades and the mitochondrial pathway. The model provides quantitative insight into how TGF-β-induced apoptosis can be modulated and regulated. Model analysis shows that upon TGF-β stimulation, the role of the TAK1-MKK3/6-p38 cascade-driven apoptosis pathway in executing cell death is predominant over Smad-mediated or michochondrial mechanisms. The model also predicts possible dynamic behavior of the TGF-β-driven apoptosis signaling in cancerous cells. Of particular importance is the model prediction that a significant reduction in the level of functional TGF-β receptors alone is insufficient for cancer cells to avoid cell death, suggesting that additional abnormal alterations in other intracellular mechanisms such as (i) downregulation of caspases, (ii) upregulation of caspase inhibitors, and/or (iii) downregulation of upstream p38 MAPK pathway effectors, may need to occur *in tandem* for cancer cells to evade apoptosis successfully. Finally, the model indicates that the ARTS pathway may play an essential role in suppressing the anti-apoptotic effect of XIAP, a potent caspase inhibitor, leading us to propose the ARTS pathway as a promising therapeutic option.

[1] Chung, S., Miles, F., Sikes, R., Cooper, C., Farach-Carson, M. and Ogunnaike, B. (2009).

Quantitative modeling and analysis of the transforming growth factor β signaling pathway. Biophys J 96, 1733-1750.

See more of this Session: In Silico Systems Biology: Intracellular Signaling and Gene Regulation
See more of this Group/Topical: Topical A: Systems Biology

(653g) Feedback Effects of Extracellular Matrix Composition On AMPK Signaling Pathway: Pathway to Novel Drug Therapeutics Targeting Atherosclerosis

Prashanthi Vandrangi[1], John J. Y. Shyy[2] and Victor G. J. Rodgers[1], (1)Bioengineering, University of California, Riverside, Riverside, CA, (2)Biomedical Science, University of California, Riverside, Riverside, CA

Table 1. The Table summarizes biomarkers that characterize atheroprotective, atherogenic, and atherosclerotic stages of blood vessels.

Table 1. The Table summarizes biomarkers that characterize atheroprotective, atherogenic, and atherosclerotic stages of blood vessels.

Stages of blood vessel	*Atheroprotective*	*Atherogenic/ Artheroprone*	*Atherosclerosis*
Characteristic feature	Three layered vessel	Inflammation Fatty streak	Extracellular calcification
Vascular smooth muscle cells	Spindle shaped, Confluent	Migration, Proliferation	Abundance
ECM configuration	Collagen	Fibronectin	Fibrinogen
Integrins	$\alpha_1\beta_1$, $\alpha_2\beta_1$	$\alpha_v\beta_3$, $\alpha_5\beta_1$	$\alpha_{IIb}\beta_3$
Shear stress at TI	5-20 dynes/cm^2	<5 dynes/cm^2	
Velocity of blood	Sustained Laminar	Turbulent	
Viscosity of blood	6.72 cP	7.24 cP	7.79 cP
[NO]	0.012 nM	↓	↓
[Ca^{2+}]	2.2 mM	↓	↓

Atheroma is initiated andprogressed in the unhealthy and artheroprone region of the blood vessel. During the process, the collagenoussubendothelium (in the command of signaling pathways) is converted to afibrillar subendothelium. This process involves the deposition of fibronectinand fibrinogen which alters the release of NO and the compliance of the vesselwall.[2] Appropriate ECM composition establishes the functioning ofthe endothelial signaling pathways in the vascular intima. In this work, we useimmunocytochemical methods that employ the distinct variations in extracellularmatrix to study the characteristic behavior of endothelial cells inatheroprotective and atherogenic stages as shown in Table 1. We hypothesize that owing to different ECMconfiguration, AMP modified protein kinase (AMPK) phosphorylation will varywith flow.

Figure 1. The Figure shows the correlation of blood flow parameters at the endothelial cell monolayer to levels of integrin pair configurations, extracellular matrix, and AMPK signaling pathway in an endothelial cell.

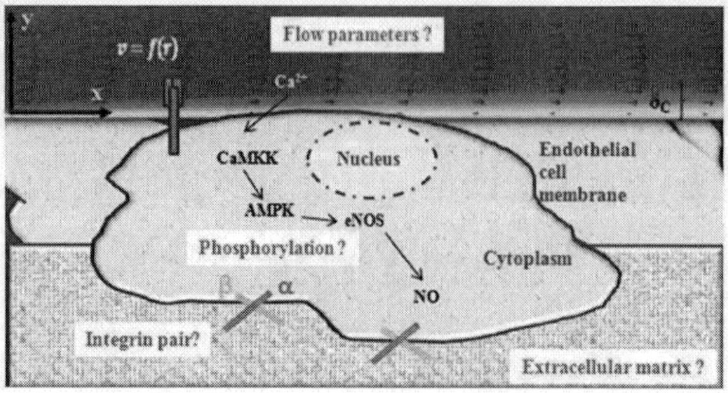

Figure 1. The Figure shows the correlation of blood flow parameters at the endothelial cell monolayer to levels of integrin pair configurations, extracellular matrix, and AMPK signaling pathway in an endothelial cell.

For each experiment, the sterilized flowcircuit is maintained at 37C, ventilated with 95% humidified air - 5% CO_2,and connected to the peristaltic pump. A parallel-plate flow channel is used toimpose laminar flow at 12 dyne/cm^2 for time intervals on confluentmonolayer of endothelial cells (ECs). The contribution of extracellular matrix for atheroprone,atherogenic, and atherosclerotic stages of the vasculature and their effects onAMPK signaling pathway is explored through these *in-vitro* experiments (Figure 1).

Exploring the stress-related AMPK signaling pathway providesstriking insights into the onset and progression of atherosclerosis. A possiblefeedback response between NO production and ECM is identified.

Keywords: *Atherosclerosis, Extracellular matrix, Signalingpathway*

References:

[1] Shyy JYJ.Extracellular Matrix Differentiating Good Flow versus Bad Flow. CirculationResearch 2009;104:931-932.

[2] Hahn C, OrrAW, Sanders JM, Jhaveri KA, and Schwartz MA. The Subendothelial ExtracellularMatrix Modulates JNK Activation by Flow. Circulation Research2009;104:995-U180.

[3] Stenman S, Vonsmitten K, and VaheriA. Fibronectin and Atherosclerosis. Acta Medica Scandinavica 1980:165-170.

[4] Levenson J, Giral P, Razavian M, Gariepy J, and SimonA. Fibrinogen and Silent Atherosclerosis in Subjects with CardiovascularRisk-Factors. Arteriosclerosis Thrombosis and Vascular Biology1995;15:1263-1268.

See more of this Session: In Silico Systems Biology: Intracellular Signaling and Gene Regulation
See more of this Group/Topical: Topical A: Systems Biology

(692a) Genome Alignment and Pair-Wise Comparison of Microbial Metabolic Networks Identifies Novel Potential Drug Targets

Joshua J. Hamilton and Jennifer L. Reed, Department of Chemical and Biological Engineering, University of Wisconsin-Madison, Madison, WI

To date, genome-scale metabolic reconstructions have been developed for over 100 prokaryotic organisms, and this number is rapidly increasing. These reconstructions serve as the foundation for constraint-based models which can be used to study cellular behavior. While a variety of methods have been developed to understand and improve cellular phenotypes, these constraint-based methods are generally only applied to models of single organisms. Recent studies have compared multiple organisms via their metabolic networks, by mapping metabolic compounds and reactions across the networks and then looking at differences and similarities in reaction and gene content (1, 2). This reaction comparison approach can be time-consuming when different nomenclatures or abbreviations are used to describe metabolites and reactions. Additionally, while a reaction alignment approach can identify network differences, it does not identify the effect of these differences on the functional states of the network. For example, one model may include a reaction as reversible while another model includes it as irreversible, but the impact of this directionality difference is generally not known. Furthermore, differences at the gene level, such as the presence of isozymes or multi-subunit enzymes, will not be identified by comparing reaction contents.

To overcome these weaknesses, we have developed a mixed integer programming approach to identify the functional differences between organisms by aligning models at the gene level. The approach first defines a set of orthologous genes based on genome sequence, and then identifies conditions under which genetic differences give rise to differences in metabolic capabilities. Because a critical issue in this analysis is the assignment of genetic orthologs, we also assessed the accuracy of existing ortholog prediction tools with respect to model annotations, and identified strengths and weaknesses of a number of these tools. We applied the algorithm to existing models of two human pathogens, *Mycobacterium tuberculosis* and *Staphylococcus aureus*, in order to explore differences in pathogenicity and drug resistance based on differences in reaction and gene content. By seeking genetic perturbation strategies (i.e. ortholog deletions) that are lethal in only one organism, we were able to identify differences in their metabolic networks (e.g. gene and reaction differences) and network capabilities (e.g. unique metabolic transformations) which point to unique metabolic functions as possible targets for new antimicrobials. This general gene-centric mixed integer programming approach allows for a rapid comparison of metabolic models based on genomic data without requiring a detailed alignment of metabolic networks.

References

(1) Thiele, et al. "A community effort towards a knowledge-base and mathematical model of the human pathogen Salmonella Typhimurium LT2." BMC Systems Biology 2011. 5(8).

(2) Oberhardt, et al. "Reconciliation of Genome-Scale Metabolic Reconstructions for Comparative Systems Analysis." PLoS Computational Biology 2011. 7(3):e1001116.

See more of this Session: Genomic Approaches to Systems Biology
See more of this Group/Topical: Topical A: Systems Biology

(692b) Insights From High-Throughput Reconstruction and Analysis of 3500 Genome-Scale Metabolic Models

Christopher Henry[1], Sam Seaver[1], Fangfang Xia[1], Scott Devoid[1], Matt DeJongh[2], Aaron Best[3], Veronika Vonstein[4], Ross Overbeek[4] and Rick Stevens[1], (1)Mathematics and Computer Science, Argonne National Laboratory, Argonne, IL, (2)Computer Science Department, Hope College, Holland, MI, (3)Biology Department, Hope College, Holland, MI, (4)Fellowship for Interpretation of Genomes, Burr Ridge, IL

As a result of next generation sequencing technology, exciting new opportunities are emerging for the application of model-based comparative approaches to improving our understanding of biological systems through the high-throughput analysis of genomic data. Here we present our work in the application of the Model SEED framework (http://seed-viewer.theseed.org/models/) [1] to construct draft genome-scale metabolic models for approximately 3500 complete genome sequences. In the reconstruction of these models, we develop new optimization-based algorithms for identifying and filling gaps in the metabolic network with increased accuracy over previous methods; we propose a new approach for predicting essential biomass precursors in microbes based on genomic evidence; and we explore how metabolic modeling may be applied to assign confidence to gene annotations and identify incorrect annotations that should be reassigned.

In the process of generating and gap-filling our 3500 draft models, we apply comparative genomics approaches to identify candidate genes that may be associated with the gapfilled reactions in each model. This includes the use of chromosomal clustering and functional co-occurrence in addition to BLAST scores to identify new high-confidence annotations that may be experimentally validated. This work has resulted in improved, consistent subsystems-based annotations for all 3500 genomes analyzed. All models and annotations are available for download from the SEED (http://pubseed.theseed.org/) and Model SEED sites.

We analyze how the properties and behavior of our 3500 draft models are conserved across the diverse set of microbial genomes included in the study. We identify the least and most variable metabolic pathways; we identify significant variability in the predicted essential metabolic genes and the redundancy of essential metabolic functions; and we explore metabolic properties that are most conserved amongst phylogenetically close organisms. This work reveals insights into the diversity of microbial genomes, the completeness of our knowledge of these genomes, and the areas of our knowledge where more gaps presently exist.

1. Henry, C.S., et al., *High-throughput generation, optimization, and analysis of genome-scale metabolic models.* Nature Biotechnology, 2010. **Nbt.1672**: p. 1-6.

(692c) Linking Genotype and Phenotype Through a Graded Dominant Mutant Phenotype Approach

Amanda M. Lanza[1], John Blazeck[2], Nathan Crook[3] and Hal Alper[1], (1)Chemical Engineering, The University of Texas at Austin, Austin, TX, (2)Department of Chemical Engineering, University of Texas at Austin, Austin, TX, (3)Chemical Engineering, University of Texas at Austin, Austin, TX

Systems biology seeks to establish a relationship between genotype and phenotype. As a result, it is necessary to accurately determine causative linkages between genes and function. Moreover, understanding global gene regulation has significant applications in genetics, biotechnology, and disease treatment. However, commonly invoked approaches for probing this regulation (such as gene deletions) do not accurately link protein function and phenotype in a causative manner. Here we describe a novel approach to extract quantitative, causative links between specific protein function and downstream gene expression. This approach is especially useful for multifunctional proteins. Specifically, we develop an approach termed "graded dominant mutant phenoytpes" in which a dominant allele is used to competitively inhibit and grade a specific protein function. Using the yeast histone acetyltransferase Gcn5p as a case study, we demonstrate the utility of this approach by identifying previously unknown gene targets and interactions and implicating a strong role for Gcn5p-acetylation in global gene repression. Our results demonstrate that traditional genetic approaches (can significantly misrepresent the number and identity of gene interactions and phenotypes. Thus, this approach serves as a novel genetic approach for systems biology.

See more of this Session: Genomic Approaches to Systems Biology
See more of this Group/Topical: Topical A: Systems Biology

(692d) Comparative Transcriptome Analysis In Baby Hamster Kidney Cells Through High Throughput Sequencing

Nitya M. Jacob[1], **Kathryn C. Johnson**[1], Nandita Vishwanathan[1], Karthik P. Jayapal[2], Chetan T. Goudar[2] and Wei-Shou Hu[1], (1)Department of Chemical Engineering and Materials Science, University of Minnesota, Minneapolis, MN, (2)Global Biological Development, Bayer HealthCare Pharmaceuticals, Berkeley, CA

Baby Hamster Kidney (BHK) cells, a widely used industrial cell line for recombinant protein production, still suffer from limitations in the availability of genomic resources in this information-rich age. We therefore applied high throughput sequencing to characterize and quantify transcript levels globally across six cDNA libraries prepared from Syrian hamster tissues as well as treated and untreated recombinant and parental BHK cell lines. Comparative analysis with the Chinese hamster revealed a high degree of homology, according the use of the Chinese hamster genome as backbone for an efficient transcriptome assembly. Overall, this effort aims to achieve a near-complete repertoire of annotated, close-to-full-length genes, spanning a range of functional classes.

Transcript abundance levels, determined by alignment and frequency computation for each library against assembled transcripts, reveal a wide dynamic range spanning nearly five orders of magnitude. Visualization in a pathway context enables one to conjecture possible physiologically active pathways in these cultured cells. Comparative transcriptome analyses between the different libraries provide insight of perturbed physiological functions under chemical treatments, including the demethylation drug 5'azacytidine. The diversity of libraries sequenced in this study also allows the exploration of polymorphisms in transcripts across different cell line and tissue sources. Furthermore, comparative analysis to the Chinese hamster genome permits the examination of sequence orthology, and the variations in codon usage between the two species, especially for those functional classes important for recombinant protein production.

These genomic resources will provide valuable process diagnostic tools, providing further opportunities to fingerprint, engineer and enhance BHK cells in their role as recombinant protein producers.

See more of this Session: Genomic Approaches to Systems Biology
See more of this Group/Topical: Topical A: Systems Biology

(692e) Transcription-Based Functional Modeling of the Host Response to Influenza Virus Infection: Defining a Healthy Immune Response

Jason E. Shoemaker[1], Satoshi Fukuyama[1], Saori Sakabe[1], Hiroaki Kitano[2] and Yoshihiro Kawaoka[1], (1)Japan Science and Technology Agency, Tokyo, Japan, (2)Systems Biology Institute, Tokyo, Japan

Transcription-based Functional Modeling of the Host Response to Influenza Virus Infection: Defining a Healthy Immune Response

Jason E Shoemaker, Satoshi Fukuyama, Saori Sakabe, Hiroaki Kitano, and Yoshihiro Kawaoka

Recent evidence suggests that an overly ambitious immune response may be the primary determinant of pathology during an influenza virus infection[1], yet the specific host functions involved, the timing and the magnitude of these host response functions invoked during a healthy immune response remains poorly defined. Thus, we collected microarray samples from the lungs of mice infected with a seasonal infection (KUKT4), the new Pandemic Swine flu (SOIV) or a mouse adapted Avian influenza (VN1203) with each strain representing an increasingly pathogenic infection. The microarray data was clustered using the Weighted Gene Coregulation Network algorithm (WGCNA). Each resulting cluster was analyzed to determine, first, if the cluster was highly enriched in a specific biological function, and, secondly, which transcription factors primarily regulate the genes within the cluster. This resulted in 14 clusters with clear functional roles and 19 transcription factors, several known to be canonical to influenza RNA sensing and immune activation, which can regulate said clusters. Finally, a linear model was constructed to relate transcription factor activity to the functional outcome of an influenza virus infection, and the model has been validated for its ability to predict functional behavior for additional influenza strains.

Yearly, seasonal influenza results in approximately 30,000 deaths and $87 billion dollars in associated economic losses in the United States [2]. There is a strong concern in the influenza community that a highly pathogenic Avian influenza, which has a 66% mortality rate [3], may become human to human transmissible and result in a pandemic comparable to the 1918 Spanish influenza during which over 50 million people lost their lives [4]. To best prepare for future pandemics, strain independent therapies must be designed to manage the infection by targeting the host machinery usurped by the virus. Previous studies have shown that selective targeting of interferon stimulated signaling pathways can effectively limit inflammation and mortality during highly pathogenic infections [1]. Thus, we first sought to identify which host functions are differentially regulated during infection and which of those functions are strain dependent.

Clustering determined which genes had highly similar regulation, and, via the "eigengene" for each cluster, also allowed for rapid visualization of strain dependent transcriptional behavior. The WGCNA package [5] was applied to microarray data collected at 14 time points from KUTK4, SOIV, VN1203 or mock infected mice. Given the large numbers of comparisons (42 conditions for ANOVA analysis when comparing against mock data), standard approaches to analyze differential expression failed to minimize the data to an interpretable size. Clustering, on the other hand, resulted in 30 statistically unique clusters, 14 of which were highly enriched in a variety of biological processes. Several of the biological functions which emerged have a clear relationship with the immune response (Toll-like Receptor signaling, inflammation, B cell/T cell activation), while others may represent host defense mechanisms whose activation was strain dependent (suppression of mRNA processing). Since the genes within each cluster are highly correlated, the first eigenvector, i.e. the eigengene, of the transcriptional response provides rapid means of visualizing gene behavior within the cluster. In all, clustering allowed us to define the timing of activation, the duration and the apparent magnitude of 14 biological functions.

Having summarized the microarray data to 14 biological functions, we sought to identify which transcription factors may be regulating each cluster. The Genbank association numbers were loaded into the GATHER tool which combines promoter sequence analysis and gene set enrichment to identify the top transcription factors. The most enriched transcription factor promoter sequences belong to interferon response factor 7 (IRF7), MAZ (a well known inflammation-related protein), NFκB and several other immune signaling proteins.

Taking the highly enriched transcription factors from the GATHER analysis, we fitted a linear model to the transcriptional data so that functional regulation during an infection could be predicted by the behavior of the 19 top transcription factors. All model construction and fitting was performed in MatLab in which the eigengene for each cluster was used as the model output while the transcription factors' transcriptional profile acted as the inputs. Initially, model parameterization was performed by fitting the responses to the moderately pathogenic infection (SOIV) and tested for its ability to reproduce functionality in the low and high pathogenic infections (KUTK4 and VN1203, respectively). In terms of the timing and magnitude of eigengene for each biological function, the model predictions were very accurate. We then improved predictive capability by parameterizing the model for data from all three infections, so that behaviors representing a low, medium and highly pathogenic infections would be included. The new model was validated against data from mice infected with different high-pathogenic Avian strain, VD5. The model well predicted gene/functional behavior and identifies several relationships between immune cell signatures which differs greatly depending on the degree of pathology for a given infection.

In conclusion, a mathematical model was constructed to predict from transcription factor data the genetic behavior at the functional level of lung tissue infected with influenza virus. The cluster results were surprisingly robust to the parameters applied, and, using the cluster eigengenes, allowed for simple temporal description of the functional behaviors differentially regulated during influenza infection. The model itself offers multiple opportunities to explore the host response to determine which biological functions may be suitable targets for therapeutic intervention. Furthermore, temporal aspects of immune activation can be explored to determine when particular immune events occur, and if delays or advances of particular biological functions may be associated with enhanced pathology.

1. Peiris, J.S., et al., *Innate immune responses to influenza A H5N1: friend or foe?* Trends Immunol, 2009. **30**(12): p. 574-84.

2. Molinari, N.A., et al., *The annual impact of seasonal influenza in the US: measuring disease burden and costs.* Vaccine, 2007. **25**(27): p. 5086-96.

3. Kobasa, D., et al., *Aberrant innate immune response in lethal infection of macaques with the 1918 influenza virus.* Nature, 2007. **445**(7125): p. 319-23.

4. Tan, S.L., et al., *Systems biology and the host response to viral infection.* Nat Biotechnol, 2007. **25**(12): p. 1383-9.

5. Langfelder, P. and S. Horvath, *WGCNA: an R package for weighted correlation network analysis.* BMC Bioinformatics, 2008. **9**: p. 559.

See more of this Session: Genomic Approaches to Systems Biology
See more of this Group/Topical: Topical A: Systems Biology

(692f) Genome-Level Quantification of the Population Dynamics of Virus and Virus-Like Defective Interfering Particles

Collin M. Timm, Fulya Akpinar and John Yin, Chemical and Biological Engineering, University of Wisconsin-Madison, Madison, WI

Viruses carry genes that can reprogram an infected cell to produce hundreds or thousands of progeny virus particles. Populations of progeny particles include both infectious and non-infectious particles. One class of non-infectious particles carry genomes that contain one or more deletions in essential viral functions, but they retain promoter sequences that are recognized by the viral polymerase. The genomes of such defective particles are able to replicate when defective and infectious virus particles co-infect the same cell. The resulting competition between defective and intact virus genomes for viral and cellular resources interferes with the normal production of infectious virus progeny. Defective interfering (DI) particles have been identified for virtually all viruses. However, open questions remain on how DI particles emerge from virus-cell interactions, how their genomic structure impacts their ability to interfere with virus infections, the role of DI particles in the pathogenesis of viral diseases such as AIDS, influenza, and hepatitis C, and the extent to which DI particles may be engineered for biomedical applications. To begin to address these questions we study a model RNA virus, vesicular stomatitis virus (VSV) and its production of virus and DI particles when infecting baby hamster kidney (BHK21) cells. Here, we employed Illumina sequencing and quantitative RT-PCR to characterize the population dynamics of DI particles relative to infectious virus particles across multiple co-infection cycles. By combining our genomic analysis with independent measures of biological activity of DI and infectious particles we propose a quantitative and mechanistic perspective of co-infection and interference. These results suggest ways DI particles may impact the dynamics of natural infections and open avenues for new anti-viral strategies based on the engineering of DI particles.

See more of this Session: Genomic Approaches to Systems Biology
See more of this Group/Topical: Topical A: Systems Biology

(692g) Identification of a Novel Target for Breast Cancer by Exploring Gene Switches On a Genome Scale

Ming Wu, Computer Science and Engineering, Michigan State University, East Lansing, MI, Li Liu, Microbiology and Molecular Genetics, Michigan State University, East Lansing, MI and Christina Chan, Chemical Engineering and Materials Science, Michigan State Uiversity, East Lansing, MI

An important dynamic feature that emerges from analyzing gene regulatory network is the "switch-like behavior" or "bistability", which is a dynamic feature of a particular gene to preferentially toggle between two steady-states. The state of gene switches play pivotal roles in cell fate decision, but identifying switches has been difficult. Therefore a challenge confronting the field is to systematically identify gene switches. We propose a top-down mining approach to identify gene switches from microarray gene expression data. Taking advantage of the tremendous amount of expression data, our approach aims to identify bimodality, which we hypothesize is an essential characteristic of a gene switch. We perform theoretical analysis based on kinetic modeling and simulation, and provide proof-of-concept applications on both synthetic and yeast microarray datasets.

Since the state of gene switches in the genetic network governs the phenotype, we postulate that recognizing specific gene switches will enable one to identify biomarkers or molecular signatures that would be better drug targets for treating a disease. We demonstrate the utility of our mining approach with human breast cancer by analyzing a paired breast cancer/normal tissue expression dataset against the integrated human gene expression dataset. Our discovery highlights TACSTD2 as an important biomarker for both ER+ and ER– breast cancer subtypes, as well as an attractive candidate for drug therapy against the triple negative (ER–, PR– (progesterone receptor) and HER2–) subtype of breast cancer. In addition, the target has potential implications for treating drug-resistant cases that are non-responsive to ER/HER2-targeted therapies. These results demonstrate the ability of our mining approach to identify candidate biomarkers and novel therapeutic targets in cancer. We predict through annotation, sequence matching of TF sites, and TF activity estimation, a novel transcriptional mechanisms by which TACSTD2 is regulated. Our experiments in MCF-7 (ER+/PR+) and MDA-MB-231 (ER–/PR–) breast cancer cell lines confirm the functional role of this gene in breast cancer and for the first time identify a likely transcription factor involved.

See more of this Session: Genomic Approaches to Systems Biology
See more of this Group/Topical: Topical A: Systems Biology

(692h) Short-Term Hepatic Gene Expression Profiling Following Thermal Injury

Qian Yang[1], Mehmet A. Orman[2], Francois Berthiaume[3], Marianthi G. Ierapetritou[4] and Ioannis (Yannis) P. Androulakis[1], (1)Department of Chemical and Biochemical Engineering, Rutgers - The State University of New Jersey, Piscataway, NJ, (2)Department of Chemical and Biochemical Engineering, Rutgers University, Piscataway, NJ, (3)Department of Biomedical Engineering, Rutgers University, Piscataway, NJ, (4)Department of Chemical and Biochemical Engineering, Rutgers, The State University of New Jersey, Piscataway, NJ

Thermal injury triggers both local and systemic responses including cellular protection mechanisms, inflammation, hypermetabolism, prolonged catabolism, organ dysfunction and immuno-suppression (1). Burn injury has a significant impact on the liver since it is the main organ controlling inflammatory response, metabolites levels and acute phase proteins production (2). It is known that inflammatory mediators triggered by cutaneous burn injury in the circulation results in persistent alterations in gene expression levels in the liver (1, 3). Therefore, liver is one of the target organs to understand the underlying molecular mechanisms of the disease state and to propose therapeutic approaches. The aim of this study is to gain a better understanding of how burn injury affects the hepatic gene expression patterns during the first 24 h following the burn injury. Gene expression patterns in both burn and shamburn (control) rat liver 24hr post-treatment, while bioinformatics tools reveal dynamic changes of gene expression profiles of burn group compared to those of control group.

A systemic hypermetabolic response was induced by applying a full-thickness burn on an area of the dorsal skin corresponding to 20% of the total body surface area (TBSA). Rats were first randomized into two groups: burn and sham burn (control group). Then, they were anesthetized (80 to 100 mg/kg ketamine + 12 to 10 mg/kg xylazine). The animal's back after removing all hair was immersed in water at 100°C for 10 s to produce a full-thickness scald injury. Immediately after burns, the animals were resuscitated with 50 mL/kg of saline injected intraperitoneally. Negative controls (sham burn) consisted of animals treated identically but immersed in warm water (37°C). Rats were single caged after burn or sham burn and given standard rat chow and water ad libitum until sacrifice. No post-burn analgesics were administered, consistent with other studies with this full thickness burn model since the nerve endings in the skin are destroyed and the skin becomes insensate (5). Animal body weights were monitored daily and found to increase at the same rate in both groups. Animals are sacrificed (starting at 9am) at different time points (0, 2, 4, 8, 16 and 24hr post-treatment, i.e., sham burn and burn) and liver tissues were collected and frozen for microarray analysis (n=3 per time point per group). The tissues were lysed and homogenized using Trizol, and the RNAs were further purified and treated with DNase using RNeasy columns (Qiagen). Then cRNAs prepared from the RNAs of liver tissues using protocols provided by Affymetrix were utilized to hybridize Rat Genome 230 2.0 Array (GeneChip, Affymetrix) comprised of more than 31,000 probe sets.

In order to analyze concurrently sham and burn expression profiles, gene expression data analysis includes normalization, filtering for "between class temporal differential expression", combining the datasets and clustering. First, DNA chip analyzer (dChip) software was used with invariant-set normalization and perfect match (PM) model to generate expression values. Then normalized two data sets corresponding to burn and sham groups were investigated to identify the differentially expressed probesets using EDGE. This step determined a set of probesets whose expression patterns were significantly altered after the treatment. Finally concatenated data sets corresponding to differentially expressed probesets in burn and sham groups were combined to form one single matrix, which was then clustered using the approach of "consensus clustering" (6). The goal was to identify subsets of transcripts with coherent expression pattern in sham and burn. In total 621 burn-responsive probesets in 4 clusters were identified. We characterized the biological relevance of the intrinsic responses by evaluating the enrichment of the corresponding subsets by using KEGG database through ARRAYTRACK (7) as well as analyzed the functions of each individual gene (8).

Consensus clustering identifies 4 statistically dominant transcriptional clusters composed of 62, 82, 404, 73 probe sets respectively. Functional characterization elucidated that these responses are mainly associated with pro-inflammation (cluster 1), fatty acid biosynthesis (cluster 2), anti-inflammation (cluster 3), and insulin signaling pathway and amino acid metabolism (cluster 4). More specifically, Cluster 1 exhibits an early up-regulation during the first two hours following thermal injury, which is related to a pro-inflammatory response. A group of inflammatory genes are involved in this cluster including cytokine, chemokines and chemokine receptors (*Il1a, Cxcl16, Ccl11* and *Ccl9*). Cluster 2 is considered to be suppressed following burn injury which is associated mainly with fatty acid biosynthesis. The genes in this cluster are involved in biosynthesis of unsaturated fatty acid pathway (*Acot1*, *Acot2*, and *Acot3*) and synthesis of ketone bodies (*Hmgcs2*). Besides, many other genes in this specific motif are relevant to fatty acid biosynthesis or transport such as *Eci1, Pigo, cyp4b1, Adfp, Pnpla8, Pank4, Crot, Etfdh.* Cluster 3 is activated around 8h post-burn which is associated with the late anti-inflammatory response. Genes in this major temporal class are important in complement and coagulation (*C2, C4bpa, C8a, Cfh, Masp1, and Serping1*), N-Glycan biosynthesis pathway (*B4galt1, Dad1, Ddost, Dpagt1, Ganab,* and *Man1b1*), and ribosome (*Rps25, Rps2*). All these functions can be interpreted as the synthesis of acute phase proteins and important anti-inflammatory mediators. Cluster 4, is mainly related to insulin regulated metabolism, is downregulated around 24h post-burn. The genes in this cluster are involved in the insulin signaling pathway (*Gck, Irs1, Mknk2, and Trip10*), phenylalanine metabolism (*Ddc*), glycine, serine and threonine metabolism (*Bhmt*), and galactose metabolism (*Gck*).

Our results indicate that the pro-inflammatory response is rapidly up-regulated within the first 2 hours following burn injury, while the anti-inflammatory response is activated later on, around 8h post burn, which would otherwise be considered as a biphasic response starting from beginning with an early downregulation and a late upregulation if control group was assumed to be a constant value (0 time point). Thus, once the pro-inflammatory response has mounted it serves as a subsequent signal for stimulating the anti-inflammatory response to inhibit the pro-inflammatory response and drive the system back to homeostasis.

We have previously determined that the dynamic profiles representing fatty acid biosynthesis and insulin regulated metabolism as well as amino acid metabolism in sham group exhibited daily oscillation, consistent with previous observations elucidating the circadian rhythmicity in the gene expression patterns of rat liver (9). Following burn injury, it was observed that the genes in the corresponding clusters lost rhythmicity and were suppressed starting from soon after (~2h) and much later (~16h) postburn respectively compared to dynamic gene expression profiles of control group. Lipid biosynthesis associated enzymes are observed to be downregulated following the burn, implying an enhanced energy demand. Downregulation of the genes involved in insulin signaling pathway further suggests a potential mechanism to explain the onset of hypermetabolism, a well-known characteristic of burn injury with potential adverse outcome.

In conclusion, our results reveal critical gene expression pattern changes triggered by burn injury which reflect host physiological and biological alterations and provides a more comprehensive understanding of the pathophysiology of the disease state. Simultaneous analysis of both burn and sham-burn groups' expression profiles is advantageous since it enables to characterize the dynamic patterns of both groups. Moreover, it also provides a more accurate estimation of the activation time and expression pattern of a certain response.

References

1. Dasu MR, Cobb JP, Laramie JM, et al. Gene expression profiles of livers from thermally injured rats. Gene 2004;327(1):51-60.

2. Baumann H, Gauldie J. The acute phase response. Immunol Today 1994;15(2):74-80.

3. Vemula M, Berthiaume F, Jayaraman A, et al. Expression profiling analysis of the metabolic and inflammatory changes following burn injury in rats. Physiol Genomics 2004;18(1):87-98.

4. Banta S, Vemula M, Yokoyama T, et al. Contribution of gene expression to metabolic fluxes in hypermetabolic livers induced through burn injury and cecal ligation and puncture in rats. Biotechnol Bioeng 2007;97(1):118-137.

5. Valenti LM, Mathieu J, Chancerelle Y, et al. High levels of endogenous nitric oxide produced after burn injury in rats arrest activated T lymphocytes in the first G1 phase of the cell cycle and then induce their apoptosis. Exp Cell Res 2005;306(1):150-167.

6. Nguyen TT, Nowakowski RS, Androulakis IP. Unsupervised selection of highly coexpressed and noncoexpressed genes using a consensus clustering approach. Omics 2009;13(3):219-237.

7. Tong W, Cao X, Harris S, et al. ArrayTrack--supporting toxicogenomic research at the US Food and Drug Administration National Center for Toxicological Research. Environmental health perspectives 2003;111(15):1819-1826.

8. Twigger S, Lu J, Shimoyama M, et al. Rat Genome Database (RGD): mapping disease onto the genome. Nucleic Acids Res 2002;30(1):125-128.

9. Ovacik M, Sukumaran S, Almon R, et al. Circadian signatures in rat liver: from gene expression to pathways. BMC Bioinformatics 2010;11(1):540.

See more of this Session: Genomic Approaches to Systems Biology
See more of this Group/Topical: Topical A: Systems Biology

AUTHOR INDEX

Aggarwal, Shilpi 79
Akpinar, Fulya 103
Alexopoulos, Aleck 24
Alper, Hal 98
Amaral, Luis A. N. 70
Androulakis, Ioannis P. 19, 61, 105
Anesiadis, Nikolaos 5
Antoniewicz, Maciek R. 42
Aprison, Erin 70
Ashe, Hilary 69
Auchtung, Jennifer M. 48
Bansal, Loveleena 87
Barton, Paul 38
Beliaev, Alex S. 76
Benedict, Matthew Nicholas 43
Benton, Michael 40
Berla, Bertram M. 82
Berthiaume, Francois 105
Best, Aaron 97
Bhushan, Mani 74
Biliouris, Konstantinos 3
Birtwistle, Marc R. 90
Blazeck, John 98
Boedicker, James 55
Bois, Justin S. 45
Burns, Mark A. 8
Calvano, Steven E. 19, 61
Carlquist, Magnus 36
Carlson, Ross P. 44
Chan, Christina 86, 104
Chatterjee, Anushree 31, 85
Chatterjee, Manash 22
Chen, Feng 11
Chen, Yu 78
Choi, Taijung 29
Chu, Yunfei 87
Chung, Seung-Wook 91
Cluett, William R. 5
Colace, Tom 22
Cooper, Carlton R. 91
Crook, Nathan 98
Daringer, Nichole 9
Dejongh, Matt 97
Devoid, Scott 97
Diamond, Scott L. 22
Doyle, Francis J. 32
Dunny, Gary M. 85
Dutta, Abhishek 36
Eddy, James A. 59
Elvitigala, Thanura 82
Fallahi-Sichani, Mohammad 54
Farach-Carson, Mary C. 91
Fernandes, Rita Lencastre 36
Flamm, Matthew H. 22
Folsom, James P. 44
Fong, Stephen S. 5
Foteinou, Panagiota T. 32
Franz, Andre 27
Fukuyama, Satoshi 100
Garcia, Hernan 55
Gauthier, John H. 67
Gawand, Pratish 49
Gernaey, Krist V. 36
Gonnerman, Matthew C. 43, 83
Goudar, Chetan T. 99
Grammel, Hartmut 27
Grill, Stephan W. 45
Hahn, Juergen 51, 87
Hamilton, Joshua J. 95
Harris, Robin 69
Harwood, Stuart 38
Hassoun, Soha 46
Heins, Anna-Lena 36
Henry, Christopher 97
Hjortsø, Martin A. 40
Hoeffner, Kai 38
Hu, Wei-Shou 31, 85, 99
Huang, Zuyi 51
Ierapetritou, Marianthi G. 105
Jacob, Nitya M. 99
Jayapal, Karthik P. 99
Jensen, Anker D. 36
Johansen, Soren J. 36
Johnson, Brian N. 8
Johnson, Kathryn C. 99
Julicher, Frank 45
Kapoor, Ashish 41
Karimi, Iftekar A. 79
Kaunas, Roland 51
Kawaoka, Yoshihiro 100
Kaznessis, Yiannis N. 3, 85
Kenis, Paul J. A. 41
Kienle, Achim 27
Kim, Hyun Uk 35
Kim, Tae Yong 35
Kimura, Yuki 41
Kiparissides, Costas 24
Kirschner, Denise E. 54
Kitano, Hiroaki 100
Klinke, David 15
Kobayashi, Hideki 5
Kolch, Walter 90
Koneru, Murali 40
Konopka, Allan E. 76
Kulkarni, Yogesh 15
Laird, Carl D. 87
Lampo, Thomas J. 14
Lantz, Anna Eliasson 36
Lanza, Amanda M. 98
Lazzara, Matthew J. 57

AUTHOR INDEX

Leamy, Alexandra K. 13
Lee, Dong-Yup 79
Lee, Kyongbum 46
Lee, Sang Yup 1, 35
Leighty, Robert 42
Leonard, Joshua N. 9
Li, Rui 48
Lin, Xiaoxia Nina 8, 78
Linderman, Jennifer J. 54
Liu, Li 104
Long, Richard L. 18
Lowry, Stephen F. 19, 61
Lundin, Luisa 36
Lv, Na 11
Mahadevan, Radhakrishnan 5, 49
Maranas, Costas D. 82
Maurya, Mano R. 29
Mavroudis, Panteleimon D. 61
McMullen, Patrick D. 70
Metcalf, William W. 43
Milne, Caroline B. 72
Minty, Jeremy J. 8
Monast, Calixte S. 57
Moon, Tae Seok 2
Morimoto, Richard I. 70
Murphy, Taylor A. 17
Nopens, Ingmar 36
Ogunnaike, Babatunde A. 91
Olson, Luke 41
Orman, Mehmet A. 105
Overbeek, Ross 97
Page, Lawrence E. 82
Pakrasi, Himadri B. 82
Pargett, Michael S. 69
Park, Jihyang 8
Pena, Pedro V. 84
Phillips, Rob 55
Pinchuk, Grigoriy E. 76
Pohl, Phillip I. 67
Prather, Kristala 4
Price, Nathan D. 43, 59, 72, 83
Ramkrishna, Doraiswami 31, 65, 85
Rao, Christopher V. 41
Reed, Jennifer L. 76, 95
Rehner, Ruxandra 27
Rodgers, Victor G. J. 93
Ruvinsky, Ilya 70
Saha, Rajib 82
Sakabe, Saori 100
Scheff, Jeremy D. 19, 61
Seaver, Sam 97
Senger, Ryan S. 68
Sengupta, Tirthankar 74
Shoemaker, Jason E. 100
Shu, Che-Chi 31, 85
Shyy, John J. Y. 93
Sinno, Talid 22
Solomon, Kevin 4
Song, Hyun-Seob 65
Sridharan, Gautham V. 46
Srienc, Friedrich 84
Srivastava, Poonam 3
Stevens, Rick 97
Stolyar, Sergey M. 76
Styczynski, Mark P. 16
Subramaniam, Shankar 29
Sun, Jingjing 10
Taffs, Reed 44
Tartakovsky, Daniel M. 29
Tiedje, James M. 48
Timm, Collin M. 103
Tirrell, David A. 14
Truong, Frank 14
Umulis, David M. 69
Unrean, Pornkamol 84
Vandrangi, Prashanthi 93
Vermeersch, Kathleen 16
Vishwanathan, Nandita 99
Voigt, Christopher A. 2
Volzing, Katherine 3
Von Kriegsheim, Alexander 90
Vonstein, Veronika 97
Vu, Trang T. 76
Wangikar, Pramod P. 74
Wei, Dong 11
Weiss, Ron 10
Winter, Peter B. 70
Worden, R. Mark 48
Wu, Ming 86, 104
Xia, Fangfang 97
Yang, Qian 105
Yin, John 103
Yoo, Tae Hyeon 14
Young, Jamey 13, 17
Zaroff, Ted A. 8